GLACIER BAY NATIONAL PARK

A BACKCOUNTRY GUIDE TO THE GLACIERS AND BEYOND

Jim DuFresne

Chapter 12 by Ken Leghorn
Additional Research by Ed Fogels and Jeff Sloss

THE MOUNTAINEERS
SEATTLE

The Mountaineers: Organized 1906
" . . . to explore, study, preserve, and enjoy the natural beauty of the Northwest."

Published by The Mountaineers, 306 2nd Avenue West
Seattle, Washington 98119

Published simultaneously in Canada by Douglas & McIntyre Ltd.
1615 Venables St., Vancouver, B. C. V5L 2H1

Manufactured in the United States of America

Designed by Bridget Culligan
Cover design by Constance Bollen
Maps by Newell Cartographics
Photos by the author unless otherwise noted
Sketches by John Svenson
Cover photos: (Top) Kayakers at Muir Glacier, by Jeff Sloss; (lower left) hiker cross-
ing stream at Outside Coast, by Jeff Sloss; (lower right) Johns Hopkins Glacier in
the West Arm, by Jim DuFresne.

Library of Congress Cataloging in Publication Data

DuFresne, Jim.
   Glacier Bay National Park.

   Includes index.
   1. Glacier Bay National Park and Preserve (Alaska)—Guide-books.
2. Hiking—Alaska—Glacier Bay National Park and Preserve.
3. Kayak touring—Alaska—Glacier Bay National Park and Preserve.
I. Leghorn, Ken.   I. title.
F737.G5D83 1987   917.98'2   87-14294
ISBN 0-89886-132-2 (pbk.)

0   9   8   7

5   4   3   2   1

To Bonnie,
For the lesson of life,
that the ordinary is rarely acceptable.

And to Jeff and Anne,
For a rare day in July on the outside coast.

CONTENTS

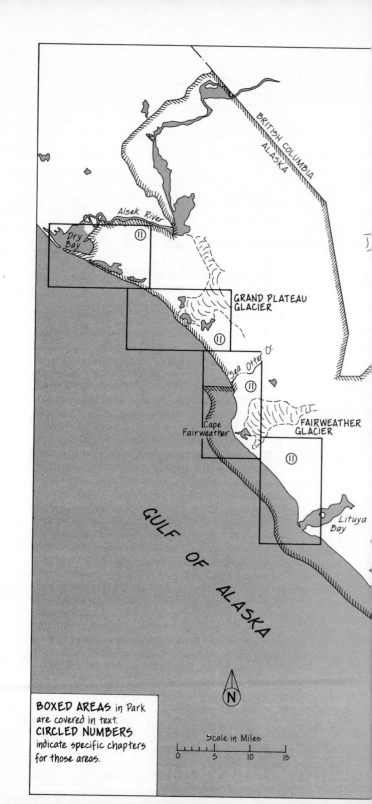

BRITISH COLUMBIA
ALASKA

Alsek River

Dry Bay

⑪

GRAND PLATEAU GLACIER

⑪

Sea Otter Cr.

⑪

Cape Fairweather

FAIRWEATHER GLACIER

⑪

Lituya Bay

GULF OF ALASKA

N

BOXED AREAS in Park are covered in text.
CIRCLED NUMBERS indicate specific chapters for those areas.

Scale in Miles
0   5   10   15

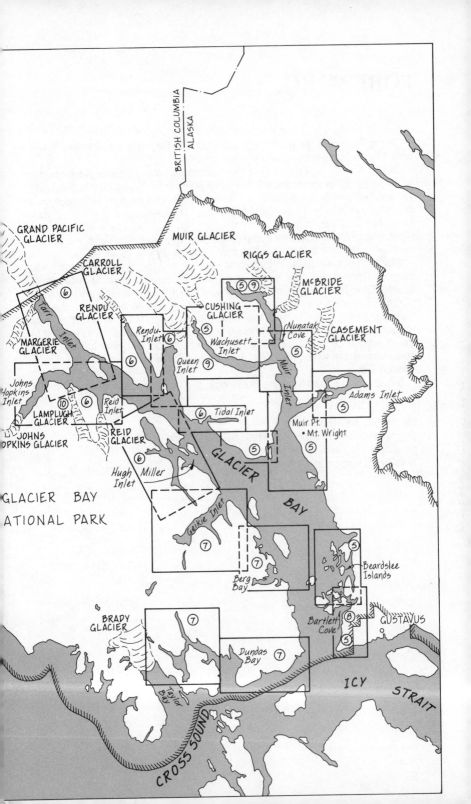

FOREWORD

As the first route guide comes out for Glacier Bay, it is important to reflect on a wilderness experience and what it means.

Over the decades Glacier Bay has become renowned for its wildness, for an opportunity to experience wilderness firsthand. The traveler's skills and knowledge are pitted against the often harsh and remote elements that make up the land and water of Glacier Bay. The sense that you are on your own, that your experiences and enjoyment are a direct result of your decisions is one element of the complex formula that makes an exciting wilderness adventure. It is this oneness that I challenge you to strive for.

Gaining knowledge is an important factor in preparing for any wilderness journey. This new book is an opportunity to expand your knowledge of Glacier Bay. It should be of value as you make your planning decisions.

But I challenge you to go beyond the route guide, to experience the wilderness that exists beyond the descriptions in print. Find what is truly unique in the area, what is unknown. Bring Glacier Bay into your life. Add to the mystique of the area. Through this effort you will hopefully find the cornerstone on which to build a truly special wilderness experience in Glacier Bay.

Explore and experience the greatness, the vastness, the wildness that is Glacier Bay.

Michael J. Tollefson
Superintendent, Glacier Bay National Park
February 25, 1987
A rainy afternoon at Bartlett Cove

Fairweather Mountains from the West Arm

ACKNOWLEDGMENTS

I once took a six-week solo paddle in Glacier Bay and in the middle of that journey didn't see anybody for 12 days. For an author who thrives on wilderness solitude, it was an incredible experience. For 12 days the only sign of man was an occasional floatplane overhead or a cruise ship off in the distance; for 12 days the only voice I heard was when I sang to myself.

It was one man's unexpected sabbatical from civilization, a time when you throw away the watch and move with the tides, reacquaint yourself with nature, rediscover who you are. It was 12 days without running into another person—and yet in this world of ours, it couldn't have been possible without the help of many people.

I deeply appreciate all the assistance from the National Park Service staff at Bartlett Cove, especially Superintendent Mike Tollefson, Assistant Superintendent Dave Spirtes, and Chief Naturalist Bruce Paige. I also thank all the naturalists and backcountry rangers who shared their knowledge of the park with me or found time to scribble suggestions and corrections on the manuscript.

I received much help from Bill Marchese and David Stewart of the Alaska Division of Tourism, Jim Johnson of Alaska Airlines, and the staff of the Alaska State Library in Juneau. Then there were the guides of Alaska Discovery—Ken Leghorn, Fred Hiltner, Jeff Sloss, Ed Fogels, and Hayden Kaden—who are more knowledgeable about Glacier Bay's backcountry than I can ever hope to be.

Summer after summer I was able to go off trekking in Alaska because of the faithful support of my wife, Peg, and the irreplaceable assistance of Amy Jeschawitz back at the home front. I am indebted to Loren Hoffman and Bill Mallonee for coming to my rescue when I suddenly needed a paddling partner and to John Barton and Carl Walter when I needed a Klepper kayak.

I also received much support from Lee Moyer of Pacific Water Sports. Donna DeShazo of The Mountaineers Books displayed unbelievable patience with me and my overdue manuscript; a trait, I suspect, that all publishers possess.

Most of all, I say farewell to the gang at Gustavus; to Buzzie Salgit (whose fingers work magic with a can of fiberglass repair), Jozee Archambault and her great banana bread, sweet Kara Berg, Steve Griffen, Diana "Bear" Privett, Bonnie, Sierra, and wonderdog Kootz. Who knows when our paths will cross again?

INTRODUCTION

It has been written time and time again that in Glacier Bay "the only constant is change."

Kayakers and hikers more than any other user of the national park are affected by this perpetual state of succession. Topographical maps and nautical charts no longer reflect most glaciers, no longer show the true length of many inlets. Valleys and river outwashes that were once padded by leafy, green mats of dryas are now jungles of alder turning pleasant hikes into battles with branches. Streams and channels are suddenly born, dry up, or change their course to the other side of the ravine. Avalanches momentarily open new routes up steep ridges while rising water temporarily closes off easy streamside hikes.

Glacier Bay is truly a wilderness.

The purpose of this book is to help backcountry users—people who step off and away from the park tour boats and float planes—to explore and enjoy Glacier Bay National Park. The book can be used first as a planning tool, offering the information you need for getting to and around the park and for selecting the right paddling or hiking trip for your party. Then, once you are in the park, this book can guide you to sidetrips and alternative routes that most visitors will not be aware of.

Of course, it is impossible to produce an exact picture of the park. This book doesn't attempt to do that. The treks described in these pages are but a few of many possibilities, and the descriptions are summaries, not step-by-step accounts. They are presented so that hikers know what to expect and may better prepare for them in the way of equipment, time, and frame of mind. Or not choose them at all.

The first part of the book consists of general travel and trip-planning information. Once you understand what the Glacier Bay National Park is, and what it offers, you can better organize and plan your trip. Keep in mind that many services, including kayak rentals, a room in the lodge, a place on the park tour boats, or a seat on the airline, should be reserved months in advance.

An extended trip into Glacier Bay is not a last minute fling.

Part II is a guide of the park from a kayaker's point of view. It describes the major routes that are followed by paddlers along both arms of the bay as well as in Icy Strait and Dundas Bay. These chapters also describe day hikes that are, for the most part, accessible to and enjoyed by kayakers alone. (The cost of chartering a float plane or boat for a single day hike up bay prohibits most other visitors from attempting them.)

Part III is devoted to hikes that do not normally require possession of a kayak. These trips include multiple day hikes as well as day hikes in areas that traditionally can be reached by foot, by park tour boat, or by commercial float plane.

The routes are rated easy, moderate and difficult. "Easy" designates level hikes where little or no brush is encountered. "Moderate" indicates treks where considerably more climbing is required, and thicker brush is more frequently encountered, but where the route is still reasonably open and easy to follow. "Difficult" applies to hikes that normally entail great physical exertion and perhaps a degree of hazard. A "difficult" trip might include numerous fordings of glacial streams, long stretches of alder, steep climbs, or difficult route finding. For example, the entire traverse of Cape Fairweather, from Lituya Bay to Dry Bay (Chapter 11) is rated as difficult and for experienced hikers only.

With or without this book, in the end you are on your own and must base decisions on what you encounter, careful deductions and your own natural instinct as to what is the best possible route at the present moment. The fact that a trip or route is described in this book is not a presentation that it is a safe one for your group. Nor has this book listed every hazard which may confront you. These are the risks, and the challenges, you assume when entering any remote and isolated area.

But because of Glacier Bay's ever changing state, wilderness travel through a trailless park will always be its main attraction to backcountry users. Enjoy the challenge this park offers but be aware of your own limitations and the present conditions existing when and where you are traveling. If a glacier river has altered dramatically or is now too dangerous to ford, change your plans and select another area to explore.

When you arrive at the park throw away time schedules and pre-arranged itineraries that are chiseled in stone. Enjoy Glacier Bay on its own terms.

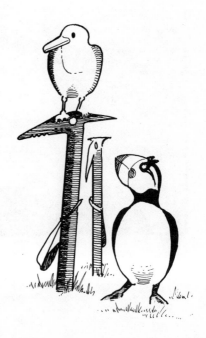

PART I

GLACIER BAY: THE NATIONAL PARK

1 THE SPECTACLE OF ICE

July 27, 1890. All . . . day it rained. The mountains were smothered in dull-colored mist and fog, the great glacier looming through the gloomy gray fog fringes with wonderful effect. It is bad weather for exploring but delightful nevertheless, making all the strange, mysterious region yet stranger and more mysterious.

(John Muir at the face of Muir Glacier)

A century later the sun was shining on Muir Glacier. Four kayakers in a pair of rigid doubles were threading their way toward the face and marveling at their luck with the weather. After a year of planning the trip, they arrived in Glacier Bay only to spend their first day in a constant drizzle moving up bay from the point where John Muir had viewed the frozen monument that now bears his name.

But on the day they reached the most famous glacier in this icy wilderness, the rain stopped, the clouds lifted, and the sky took on a deep shade of blue. They found themselves balancing between fire and ice in a struggle of contrasts. The August sun sent the temperatures soaring toward the eighties, and for the first time during the trip the kayakers greeted the warmth by stripping down to T-shirts and bare backs. But the thick icepack in front of the glacier crackled like the crushed ice in a summer drink, and kept them a half-mile away.

Even at a half-mile, the glacier was an overpowering curtain of ice suspended between two mountains; it appeared to be no more than a football field away. There was a rare beauty in the chiseled cracks and folds of its face, where every imaginable shade of blue could be found. And it was silent; only an occasional pop was heard from the surrounding icepack. The kayakers were quietly studying the natural wonder they had worked so hard to reach when it suddenly came to life for them. There was a crack, then a spill of shattered ice, followed by a huge part of the face dislodging and dropping into the inlet with an enormous explosion.

It was a feast not only for the eyes but for the ears as well. The surrounding mountains amplified the sound and filled the inlet with thunder on a cloudless day. When the newborn iceberg surfaced, it touched off a series of large swells. One by one they fanned out toward the kayakers, who watched the rolling waves raise the boats and the icepack in a gentle uplifting motion. When the last one passed by, the kayakers turned to each other and remarked on the spectacle. Then like Muir and so many others before them, they slipped back into the silence of Glacier Bay to patiently wait for the next performance in this theatre of ice.

"UNSPEAKABLY PURE AND SUBLIME"

Glacier Bay National Park and Preserve is located across Icy Strait from Chichagof Island, some 60 miles northwest of Juneau in Southeast Alaska. The bay itself, 62 miles long and 10 miles at its widest point, is surrounded by a horseshoe rim of mountains that include the Fairweather Range to the west, the Chilkat Range to the east, and the St. Elias, Alsek, and Takhinsha mountains to the north.

Icebergs in Reid Inlet in the West Arm

Although it is best known for its 16 tidewater glaciers, the park is the wondrous scene of much more than just the rivers of ice. While glaciologists study the retreat and advancement of glaciers, plant ecologists marvel at the trail of succession that the departure of glaciers initiates. You can travel the length of the park and witness the beginning of life step by step, starting up bay with the newly exposed rock and moraines that lie next to retreating ice and ending with the lush rain forest around Bartlett Cove.

Visitors who come seeking the glaciers are often stunned by the beauty of the coastal Fairweather Range, crowned by Mount Fairweather at 15,300 feet and described as some of the most spectacular glaciated mountains in the world. Equally impressive is the park's rich wildlife, especially the marine life, which includes porpoises, sea lions, seals, otters, and three species of whales. For a fortunate few every summer, Glacier Bay completely unveils itself as visitors witness two of nature's most stunning sights—the thundering performance of a glacier and the acrobatic display of a humpback whale breaching.

But more than the ice, mountains, or whales, Glacier Bay is wilderness; undeveloped beyond Bartlett Cove, unyielding to the inexperienced, and uncharted where glaciers have changed the topography faster than man can map it. This Glacier Bay belongs to the backcountry users who arrive in kayaks or on foot to explore and experience the remote corners and passageways of the 3.28-million-acre park.

Bluewater paddlers in sea-touring kayaks will find Glacier Bay especially well suited to their mode of travel. The heart of the park is the inner bay, which extends from Icy Strait north to near the Canada-U.S. border, but only after first splitting into Muir Inlet and what is commonly referred to as the West Arm. Departing from these arms are 10 more inlets that serve as the gateways to even calmer water and more spectacular glacial scenery.

The drop-off service provided by the concessionaire's tour boats, along with the usually calm conditions of the inlets up bay, allows less experienced kayakers to explore Glacier Bay without enduring open water, major crossings, or the long haul from Bartlett Cove. Those with experience and time could paddle for months and never leave the shoreline of the inner bay; or they could take on the challenge of Icy Strait to reach and explore the extensive fiords, such as Dundas Bay, that penetrate the southern boundary of the park.

In the best of Alaska's traditions, Glacier Bay is another of the state's roadless and trailless parks. The only routes are around Bartlett Cove, where there are one road and three miles of maintained trails. But hiking opportunities abound throughout the park along such natural corridors as gravel moraines, glacial outwashes, beaches, ridges, and stagnant remnants of ice left behind by retreating glaciers. Reached primarily by kayaks, most make for ideal day hikes away from the shoreline and offer pleasant breaks in the middle of a long paddle.

The exceptions are found at Wolf Point and Riggs Glacier, two places in Muir Inlet accessible to those who want nothing to do with double-bladed paddles. Here you can be dropped off by a tour boat and spend days hiking around the glacial wonders of the upper bay. The most adventurous backpacking trip within the park, however, is the challenging 70-mile trek along the outside coast from Lituya Bay to Dry Bay. Perhaps Alaska's finest wilderness beach for hiking, the outside coast is a traverse around Cape Fairweather where backpackers find themselves treading a thin strip between the pounding surf of the Pacific Ocean and the glaciated peaks of the Fairweather Range.

Like any other wilderness park, Glacier Bay should never be taken lightly. You must be totally self-sufficient and well prepared for any kind of weather or mishap when entering the backcountry.

But unlike most other parks, you will depart from this one with a new perspective on ice and the beauty it creates. Despite the park's many intriguing facets, in the end it is the glaciers, overwhelmingly beautiful, that overpower the imagination and form everlasting impressions.

John Muir was the first of many visitors affected by this natural phenomenon, and his words, even after almost a hundred years, still best describe the magnificence of Glacier Bay. After camping near Grand Pacific Glacier, Muir climbed to over 1,000 feet and for the first time during his rainy trip in 1879, the drizzle stopped and the clouds broke.

Sunshine streamed through the luminous fringes of the clouds and fell on the green waters of the fiord, the glittering bergs, the crystal bluffs of the vast glacier, the intensely white, far-reaching fields of ice, and the spiritual heights of the Fairweather Range, which [are] now hidden, now partly revealed . . . the whole making a picture of icy wilderness unspeakably pure and sublime.

2 ICE, LIFE, AND TIME

The alder had me in its tangled grasp; one branch around my arm, two around my legs, and a third that came from behind and smacked the back of my head. Cursing and sweating, I finally broke free and reached a clear spot on The Nunatak. Fifteen years ago this hill overlooking Muir Inlet was an easy romp to spectacular views. Today it is a struggle for only the strong at heart who are willing to buck the alder for a few hours in order to reach the top.

I struggled but now was enjoying my hard-earned panorama when a moose trotted into view below, heading toward the river that flowed from Red Mountain into Nunatak Cove. Moose didn't arrive in the park until the 1940s, but now they were common; this was already my third sighting during this trip and I watched it only casually.

It was the animal following the moose that made me pull out a telephoto lens. What I first thought was a calf turned out to be a wolf; suddenly I became a balcony spectator watching the high drama of nature—one animal pursuing another with life and death at stake.

The moose trotted briskly for the river and crossed it at a bend. Once on the other side, it turned and watched its pursuer. I also watched the wolf closely. I knew it was going to have problems with the swift river because I had crossed it earlier and couldn't keep my shirttails dry. The animal slowed down considerably and stalled in midstream, where it was battling the strong current. At this point the moose ran back across the river and again turned to watch its struggling opponent. When the wolf finally dragged itself out of the icy water, a truce was called. The two animals just stared at each other from opposite banks.

Then they departed in different directions and I returned to bucking the alder on The Nunatak.

THE LITTLE ICE AGE

Almost from the beginning of time, glaciers and ice have played a continuing role in shaping the world. During the Pleistocene Era, which began two to three million years ago, continental glaciation covered much of North America, carving valleys and fiords and leaving behind lakes, rivers, and fertile plains. Today glaciers and polar ice still cover 10 percent of the world, an area equal to what is being farmed, and at Antarctic and Greenland the ice caps measure 2 miles thick. There is more water within the world's ice than there is in all the lakes and rivers combined.

In this realm of glaciology, Glacier Bay stands out as a priceless laboratory. Here the ice has a history of repeating itself. The park and most of Southeast Alaska were covered and carved by glaciers during the Wisconsin Ice Age, which began 25,000 years ago as a period within the Pleistocene Era. In the warming period that followed, the glaciers pulled back, exposing such areas as Muir, Wachusett, and Adams inlets which were repopulated by lush hemlock and spruce forests for several thousand years.

But harsh climatic conditions and cold temperatures returned 4,000 years

Reid Inlet in the West Arm

ago, giving rise once again to advancing glaciers. This era is called Neoglacial and is more commonly known as the Little Ice Age. It took place around the world but nowhere as noticeably as in Glacier Bay. Here conditions were ideal for the creation and growth of glaciers. The high coastal range on the west side of the park trapped the snow and ice from the moisture-laden storms that swept across the North Pacific, building the vast Brady Icefield. The heavy accumulation of snow turned into ice, which gravity funneled down the many deep troughs and fiords that sliced into the Fairweather Range.

The glacial ice spilled out of the mountains and began another southeasterly thrust, filling the main fiord of Glacier Bay and extending into Icy Strait. A thousand years later Muir Glacier began advancing down its inlet, locking the trees into a frozen museum of interglacial wood, only to be discovered centuries later.

Saltwater is extremely destructive to ice. The huge, ancient glacier, like tidewater glaciers today, was able to advance into the bay only by building a protective shield between itself and the sea. By plucking rock debris both from the ridges and valley bottom and pushing this material ahead of it, the glacier built a shoal, actually an underwater moraine, that allowed it to keep advancing in the deep waterway.

By the early 1700s, the vast flow of ice had driven out the Tlingit Indians living and hunting in the bay. They escaped by crossing Icy Strait; on the other side they eventually established the settlement of Hoonah. The main glacier finally reached the open channel by the mid-eighteenth century. At that point the glacier was 4,000 feet thick and 20 miles wide at its terminus and extended 100 miles from the mountains where it was born. But the open waters of Icy Strait proved too destructive for the massive ice flow. The strong currents forced the glacier to retreat from its shoal and reenter deep water. No longer could the underwater moraine protect the vulnerable snout of the glacier.

Once that had happened, the ice began to retreat with amazing speed. Never in recorded history has a glacier retreated as quickly as the one that filled Glacier Bay in 1750. When Captain George Vancouver sailed through to chart the area in 1794, there was already a 6-mile inlet protruding up the bay. The glacier had retreated 48 miles by 1879 when Muir, looking for evidence to support the continental glaciation theory, made his first trip to the area. By then the glacier named after him was already separated from the main trunk in the West Arm.

By the 1890s, Glacier Bay was no longer unknown. Having been "discovered" by Muir, it was quickly promoted by travel writers who made it known throughout the world as the place where you could go from a boat to the Little Ice Age in a single bound. Harry Fielding Reid, a pioneering glaciologist, soon arrived and with his theodolite and plane table began mapping glacier positions. Steamboats loaded with tourists puffed their way up the bay on a regular basis. Muir, who had built a cabin on Muir Point near the face of his glacier, found the people to be an interesting species.

> The *Queen* arrived about 2:30 P.M. with two hundred and thirty tourists. What a show they made with their ribbons and Kodaks! All seemed happy and enthusiastic, though it was curious to see how promptly all of them ceased gazing when the dinner bell rang, and how many turned from the great crystal world of ice to look curiously at the Indians that came alongside to sell trinkets.

On September 10, 1899, an earthquake rocked the Alaskan coast, and a new chapter began in the natural history of Glacier Bay. Muir's ice front was devastated, and within hours the bay was filled with a mass of ice. Ships could get no closer than 5 miles from the face of Muir Glacier. The glaciers themselves went into rapid retreat, and by 1916 the Grand Pacific Glacier had retreated 65 miles from the bay's mouth to the head of Tarr Inlet, having exposed the front of Johns Hopkins Glacier. The rapid retreat stunned the world, and scientists poured in to document it and come up with an explanation for the diversity of glacial patterns.

What they discovered was that glaciers are complex natural phenomena, not just masses of ice. They occur wherever snowfall exceeds snowmelt and are born when weight and pressure compress snow from flakes to granular crystals and, eventually, to solid ice.

The tremendous weight of accumulating snow creates ice so hard that it possesses unique characteristics. The pressure squeezes out all cracks and air bubbles, allowing the dense ice to act as a prism, reflecting only blue wavelengths and absorbing all other colors. Along the face and the exposed sections of the glaciers, this dark blue tinge is best seen on overcast days, when more blue than other colors passes through the cloud covering. The dense glacial ice also melts at an extremely slow rate and has the ability to flow. Along its underside, where the pressure is greatest, a glacier is plastic, flowing as a ponderous, viscous river between mountains, sometimes as fast as several feet a day.

More than just ice, glaciers also contain tons of rock, boulders and gravel scoured and plucked from the mountainsides and valleys along their course. By bulldozing the rubble in front of them or carrying it on top, they recarve the typical V-shaped river valley, changing it to a U-shaped trough. Much of the rock debris ends up as lateral moraines, mounds and ridges shoved up by a glacier along its sides. Lateral moraines of a tributary glacier often become medial moraines down the main trunk of ice. They appear as ribbonlike strips of rubble after being squeezed together into the main glacier.

On land, the retreating glaciers deposit their debris and gravel in mounds and ridges known as terminal moraines. But those whose snouts reach tidal water are far more impressive in their work. As the water undermines the faces of these tidewater glaciers, huge chunks of ice, some as large as 200 feet and weighing several tons, break loose and thunder into the sea with a backdrop of mist.

The icebergs then begin their journey toward Icy Strait, though few ever make it past Tlingit Point, where the bay's two arms join to form a single body of water. Those that are exceptionally large and deep blue are composed of denser ice and could last up to a week. Others are white, signifying trapped air bubbles and an end that is near. Blackish icebergs possibly may have come from near the bottom of the glacier. Still others look like their surface has been riveted with stones and gravel. All have intriguing shapes that constantly change and spark the imagination of those viewing them from a beach, a ship, or the seat of a kayak.

Glaciers balance between the rate snow accumulates in the ice field, its source, and the rate ice melts at the other end, its snout. When the rate of accumulation is greater, the snout of the glacier advances across the land or deeper into the bay. When the melting exceeds the accumulation, the glacier retreats toward its ice field. Muir Glacier is retreating. At one point it had withdrawn 5 miles in seven years and until recently was retreating 0.25 mile a year. It has since slowed down, but glaciologists still predict it is only a matter of years before the famous ice floe retreats totally beyond the inlet, thus ending its days as a tidewater glacier.

Despite the steady retreat of Muir Glacier, the frozen monuments of Glacier Bay are not melting into history. Of the park's 16 tidewater glaciers, six are advancing, three retreating, and eight holding their own. The advancing ice is found in the West Arm of the park, where the higher elevations and the source of Brady Icefield override the melt factor.

The most notable advance is that of Johns Hopkins Glacier, which has been steadily moving forward for the last 50 years, literally clogging its inlet with ice that stops most boats 4 to 5 miles from the face. The Grand Pacific (which by 1925 had retreated across the international border, giving British Columbia access to seawater at the head of Tarr Inlet) suddenly began advancing, reentering the United States 29 years later.

Geologists used to believe there were four major ice ages in the last million years but now prefer to view them as a series of glacial advances, retreats, and readvances, as many as 40 such movements. Some theorize that in Glacier Bay the movements are not affected by climate as much as

Johns Hopkins Glacier (Bill Mallonee photo)

they are self-sustained, localized cycles of the ice; that the ice which carved and once filled the bay has retreated for now but surely will return.

These theories suggest that the park will forever be a living laboratory where the effects and patterns of glaciers can be studied; . . . that the flora and fauna which reclaim the barren land when the glacier retreats will again be pushed out and the topography subsequently recarved; . . . that, here in Glacier Bay, time has been suspended and we have the rare opportunity of watching the creation of our world.

NEW LIFE IN AN OLD LAND

Until the 1800s, most of Glacier Bay was buried in ice, enduring a continuous winter that lasted for centuries. But when the glaciers began to retreat, change, minute and slow in our way of thinking but rapid on the world's calendar, began to take place. In many places the ice was up to a mile thick. The land, after being depressed under the enormous weight for so long, began to rise as soon as it was uncovered.

The rate of rebound is astonishing. Around Bartlett Cove the land rises approximately 1.5 inches every year. Sections up bay that only recently were uncovered by the glaciers rise even more rapidly; often kayakers discover that what appears as a waterway on their USGS (U.S. Geological Survey) topographicals is now a short portage.

When the ice melts, boulders, bare bedrock, and mounds of gravel and glacial tilt are left behind. But their existence is short-lived at best, as life tends to move in quickly. It begins with "black crust," a feltlike mat of algae that stabilizes silt and holds in moisture. It reaches a climax when the land supports a thick spruce-hemlock forest. In between is a fascinating process known as postglacial plant succession. If glaciologists were thrilled with the ice they found in the bay, plant ecologists were equally delighted when the glaciers retreated, leaving a chronology of plant succession in their wake.

When visitors take a boat ride up bay, they are treated to the transformation of life. They begin in the lush rain forests that surround Bartlett Cove and sail up Muir Inlet. They watch spruce-hemlock forests give way to shorelines of spruce and then to fields of willow and entangling alder. Farther up, they see soft mats of dryas and fireweed, then bare rock, and finally, the glacier itself. Glacier Bay plant life, from its climax to its beginnings, all in a single-day boat ride.

Each plant, in the careful order of succession, seems to have both a purpose and a time of departure. Dryas is particularly important. As a pioneer plant, it forms dense mats on sands and gravels, whose lack of nitrogen prevents most other species from taking root. Although backpackers cherish a soft dryas mat as a place to pitch a tent, the plant's real value is its ability, with the assistance of root bacteria, to take nitrogen from the air and add it to the depleted soil. It was this small greenish plant that pioneered in much of Europe and North America when the last Ice Age ended.

Sitka alder continues the process of returning nitrogen to the soil. Once a glacier has retreated, alder takes only 30 to 40 years to move in and cover what began as a barren outwash plain. In that relatively short period of time, the alder advances so quickly that it can turn a natural hiking route into an impassable thicket of branches and trunks, which the majority of backpackers would rather not contend with. It is the retreating ice and advancing alder that keeps Glacier Bay a constantly changing park as far as hikers are concerned. As the glaciers melt, new routes to the interior open up, only to be closed a few decades later by a blanket of alder 10 feet tall.

Within the thick canopy of alder and willow, cottonwoods begin growing once the soil has sufficient levels of nitrogen. Many cottonwoods eventually reach above the alder, but their dominance is only temporary. In the shady understory, the lack of light prevents seedlings of cottonwood and alder from establishing themselves and slows down the previous pattern of jungle-like growth. The existing plants prepare the land for the next step of succession as Sitka spruce begins to thrive in the shaded conditions. Each wave of plants changes the environment—whether by adding nutrients, reducing available light, or retaining moisture—and paves the way for the next community.

The spruce will continue to increase until its stand is pure and thick, shading the forest floor to prevent all but a few hardy species, such as ferns, mosses, and devil's club, from surviving. Eventually, shade-tolerant western hemlock displaces the spruce to form the final forest, the climax stage of plant succession. In Bartlett Cove, where the ice has been gone for less than 200 years, you find a forest community of spruce and hemlock competing for light, moisture, and nutrients. In other areas of the park, muskeg will represent the final stage, as it topples the towering spruce and hemlock trees and returns the land to a lush, green openness, where poor drainage retards decay and brings on a new wave of plants.

THE FAUNA

The orderly way that plants repopulate the land is not carried over into the animal kingdom. There seems to be no true pioneer species preparing the land for others; and land mammals in general face numerous obstacles in returning to glacier-freed lowlands. They must either walk or swim, not having the luxury of hitching rides with the wind, waves, or birds as seeds and spores do.

But in the sea, life flows with every incoming tide as animals find open corridors to the faces of retreating glaciers. The cold Pacific seawater, loaded with nutrients and oxygen, supports an abundance of microscopic algae and phytoplankton, the foundation of the marine food chain. They in turn support zooplankton, billions of invertebrates, and tiny crustaceans. In straits and narrow waterways, the swift currents and tidal rips push the zooplankton near the surface, where enormous schools of herring and other small species of fish feed themselves.

These fish become food for salmon, which make annual spawning runs up many of the park's rivers and streams, including the Dundas, Bartlett, Beartrack, and many rivers along the outside coast. Silver, pink, sockeye, and chum salmon spawn in the park, and from middle to late summer create a wilderness feast not only for marine life but also for protein-starved terrestrial communities. Seals feed on the returning salmon near the mouths of streams, while brown bears, black bears, wolves, coyotes, minks, ravens, and bald eagles feed along the banks or plunge into the icy water to gorge themselves as well. Up bay, glacial streams carry too much silt for spawning salmon. But in time, as alders increase, changing rocky outwashes into fields of thick brush, the streams become clearer and attract a few wayward spawners, the beginning perhaps of a frenzied run of thousands.

Sometimes the ice itself is the reason for life. The constant calving of tidewater glaciers stirs up the water and brings food near the surface and within reach of the diving kittiwakes, gulls, and arctic terns. Nearby on the icepack the most common residents are harbor seals. These mammals use the floating icebergs for places to sun, rest, and most important, to give birth in the spring. The icepack becomes the pupping ground from May through August in a trend that biologists believe developed in Glacier Bay.

When there were tidewater glaciers in Wachusett and Adams inlets, mothers and pups congregated there, but as the ice left so did the seals. Undoubtedly the greatest concentration of seals is in Johns Hopkins Inlet, where there are thousands. Kayakers paddling here or through other ice-choked fiords should never startle a mother or pup on an iceberg or approach them too close. Startled mothers often lead to separated pups that starve or become easy prey for such natural predators as killer whales.

The killer whale or orca is the smallest of the three species of whales that inhabit the park. This toothed whale is rarely longer than 30 feet and is very distinctive with its bold black-and-white color pattern. Along with the contrasting pattern found on its belly, flanks, and head, what makes the whale so recognizable is its triangular dorsal fin. On old males this fin may reach 6 feet.

Orcas often swim in packs, called pods, and when several of the tall fins knife through the water, it is an impressive sight. By swimming and hunting in teams, orcas have been known to kill blue whales, the world's largest animal. Usually, however, they feed on sea lions, seals, sharks, porpoises, and squid. They are extremely fast swimmers, often reaching speeds of 29 miles per hour.

The other two species are both baleen whales, labeled so because they filter seawater through their horny baleen plates to comb out food. The humpback is by far the largest, often reaching a weight of 45 tons; ironically, it is lured to the park by Glacier Bay's abundance of krill, a small crustacean one step above microscopic phytoplankton in the food chain. Adult humpbacks average 40 to 50 feet in length and contain 600 to 800 baleen plates in their mouths, which they are able to open 90 degrees to scoop up gallons of water containing masses of sea organisms.

Humpbacks arrive in Glacier Bay each summer to feed and build up fat that will sustain them throughout the winter when they migrate as far south as Hawaii. They are known to be acrobatic leapers and playful swimmers. What kayakers often spot is the whale's spout, which can be seen and heard at least a half-mile away. With a pair of binoculars, you can also see its small dorsal fin as it skims the surface, followed by its huge fluke or tail fin rising up in a cloud of mist and water for a thrust that will send the animal to the bay's depths. Luckier visitors will witness the whale roll on the surface, displaying its long flippers or possibly even breaching when it leaps out of the sea in an impressive display of gymnastics.

Minke whales are considerably smaller, ranging from 20 to 33 feet in length, and fast swimmers, capable of reaching speeds of 20 miles per hour. They feed on cod, herring, and pollock while in the park, and krill after they migrate to the south. Minke whales are often found near the shoreline, where their small size and white band along the fore flippers serve as their field marks.

Traditionally Glacier Bay has been an important summer habitat for whales. Controversy erupted in the summer of 1978 when some of the humpback whales in the park suddenly departed. The number of humpbacks remained low in the following years, and one researcher pointed an accusing finger at the increasing vessel traffic. Others attributed the whale departures to a reduction in available prey. Teams of scientists began an intense study of the endangered humpback, which led the National Park Service (NPS) to restrict the number of boats allowed in the bay, and place limits on their speed, and on how close they can approach a whale.

In 1984, 24 humpback whales were sighted in the park. But environmentalists warn us that although the whales have returned in force to their haven in Glacier Bay, their struggle outside the preserve continues.

On land, animals and especially large mammals struggle in a different way to repopulate areas recently freed by the glaciers. Many have shown a special means to speed up recolonization, as river otters, mink, and brown and black bears swim around ice barriers that prevent access by other land animals. The first populations that come into newly exposed sections tend to be transient, moving on to other feeding areas for part of the year. Gradually, as plants develop and salmon begin to spawn in the streams, permanent-resident populations build.

Bears range throughout the park, with black bears populating the lower

Moose rack in Hugh Miller Inlet

Mountain goats on Gloomy Knob

reaches of the bay and brownies living in the upper arms and on the outside coast. Adult black bears range in size from 180 to 250 pounds, whereas their brown cousins may weigh from 500 to 900 pounds. Bears are omnivorous and creatures of opportunity. In May and June they are often spotted along the shoreline after a long winter's nap, feeding on fresh sprouts, roots, or the carcass of any animal they come across. As summer progresses, they shift to eating salmon and berries.

Other early occupants include mountain goats and hoary marmots, whose natural habitat of alpine meadows and ridges is usually the first to develop after waning ice exposes bare ground. Mountain goats now range through most of the park's alpine areas, feeding on grasses and shrubs during the summer and dropping to lower elevations in the winter to consume willow buds, dried grasses, and occasionally the buds of a spruce. During the summer, most visitors view them from the water as white dots on a mountainside. But goats are not particularly wary, and hikers who endure the alder to reach alpine areas often can observe them from reasonably close distances.

Moose, wolves, and coyotes also live in the park but are still in the colonizing stage. Moose, which range in size from 1,000 to 1,600 pounds, are believed to have arrived in the Adams Inlet area in the 1940s after migrating through Endicott Pass from Lynn Canal. Now they are found throughout much of the eastern side of the park and have spread around Tlingit Point into Tarr Inlet as well as south to Dundas Bay.

Then there are animals that have yet to make the journey into the new land or establish a significant population. The most notable is Sitka black-tailed deer, which are abundant throughout Southeast Alaska. Others include beavers, lynx, and snowshoe hares.

Ptarmigan near Gloomy Knob

What connects the riches of the marine world with the terrestrial in Glacier Bay is the shoreline, this tidal zone where life from both the sea and the land tread. At low tide you can tiptoe through the pools and beds of kelp to see starfish, sea urchins, mussels, barnacles, and snails. Come at dusk and you may have to share the strip with a brownie digging clams or eating gooseneck barnacles. You may also be chased away by an angry arctic tern, dive-bombing you from above when you wander too close to its nesting site.

Glacier Bay boasts more than 200 species of birds that live or pass through its boundaries, and many of them are best viewed on the water or near this shoreline strip. Arctic terns, the legendary migrants that travel 20,000 miles each year back and forth from their winter range in the southern hemisphere, nest in the gravel soils of glacial outwashes as do mew gulls, killdeer, and the comical oystercatchers with their distinctive orange bills. Remote islands or rocky islets provide safe nesting sites for tufted and horned puffins, pigeon guillemots, cormorants, and herring gulls. Others, such as old squaw ducks and phalaropes, are best seen at mouths of bays or narrows where swift currents stir an abundance of shrimp and fish to the surface. Along high cliffs, rising straight above the seas, are kittiwake colonies. Canada geese flock to the tidal flats of Adams Inlet.

Birds, with the ability to fly, are generally thought of as one of the first to populate the new land, bringing with them seeds for glacial tilt and crustaceans to newly formed ponds. In fact, many birds bound for the arctic or Aleutian Islands stop in Glacier Bay instead because the pioneer plant communities of the upper regions of the bay are so similar. But even here, change is inevitable. As the plant communities succeed to the spruce-hemlock stage, these birds will move on and be replaced by forest dwellers such as thrushes, woodpeckers, and grouse.

3 GETTING TO AND AROUND GLACIER BAY

Three hikers sat on a water-soaked log near the mouth of Lituya Bay, peering across at the other side. Every now and then the forested shoreline briefly appeared, then more clouds and mist would roll in from the Pacific Ocean and obscure the scene. The hikers' pick-up was scheduled for 4 P.M.; it was now 5:10 but there was never much hope of the small bush plane reaching them.

After seven clear days on the outside coast of Glacier Bay, their luck had run out. They woke up on their final morning weathered in with clouds hanging so low you could snatch them from the sky. The hikers threw their packs together and trooped down to the beach to wait for their pick-up, but it was merely ritual. They left the tent standing and passed the time by doing an inventory of their remaining supplies. It was going to be a cup of soup and granola bars for dinner, and another soggy night in a soaked tent and damp sleeping bags.

The weather's only blessing was that it made leaving the wilderness much easier. Another sunny day on the outside coast and the hikers would have been heartsick at having to step into the plane and leave the Fairweather

Floatplane pick-up on Reid Inlet (Bill Mallonee photo)

Flight back from Reid Inlet (Bill Mallonee photo)

Range and the pounding surf behind. Now they dreamed of the soft beds, real food, and hot sauna waiting for them in Gustavus.

Somewhere in the distance a low buzzing sounded. One of the hikers didn't catch it at first, as this was her first trip to Alaska and she hadn't learned to isolate that distinct sound. The other two, veteran travelers of Alaska's wilderness, picked up on it immediately. They were cautious, of course, discussing how the plane was probably on its way home to a cabin on Dry Bay.

But as the sound grew louder, they could not contain their anxiety and found themselves standing at the water's edge and searching the cloudy mist south of them. Of all the panoramas and scenic views of the outside coast provided them, none could have been more dramatic that final day than the sight of the small plane emerging from the clouds and dipping its wing above them. They cheered, they clapped, and then they ran off into the wet bush to take down the tent. Sometimes leaving the wilderness can be as invigorating as entering it.

GETTING TO GLACIER BAY

Traveling to Glacier Bay is more complicated than going to most national parks, as it can be reached only by plane or boat. Although a 10-mile road connects Bartlett Cove with Gustavus, the park and the small service center are isolated from the Alaska highway system, prohibiting visitors from driving there. Nor is the park or Gustavus a port-of-call for the Alaska Marine Highway System's vessels, as Hoonah, across Icy Strait, is the closest stop for the state-maintained ferry system.

The quickest and most popular way to enter the park is Alaska Airlines' daily jet service to Gustavus during the summer. The 10-minute flight from Juneau leaves for Gustavus in early evening and returns immediately. For information on schedules and fares, call Alaska Airlines toll free at 800 426-0333; or call 907 697-2203 in Gustavus or 907 789-0600 in Juneau.

Boat service between the two towns is offered during the summer by Glacier Bay Yacht Tours aboard its vessel *Glacier Express*. The boat departs from Juneau for Gustavus in early morning (daily, except Tuesday) for the 3-hour trip. For more information, contact Glacier Bay Yacht Tours, 79 Egan Drive, Suite 10, Juneau, AK, 99802; call 907 586-6883.

Glacier Bay can also be reached by a number of air taxi operators from Juneau, Hoonah, Haines, Sitka, and Yakutat. Many have regularly scheduled flights and per-seat fares to Gustavus. Parties of three or four might find it cheaper and more convenient to charter a plane straight to Bartlett Cove or their drop-off point within the park's backcountry. Check the appendix for a list of air taxi operators.

A bus meets all flights and provides transportation between the airport and Bartlett Cove at flight time only. For taxi service at other times, call the Glacier Bay Lodge at 697-2225.

GETTING AROUND GLACIER BAY

Every summer a few visitors arrive at Bartlett Cove, look around, and ask where the glaciers are. After finally reaching this isolated park, they are dismayed to find out they're still 43 miles away from the nearest glacier and it will take a boat ride, and the purchase of yet another ticket, to view it.

When planning a trip to Glacier Bay, you should think of it not as a national park but as a marine park; indeed, the majority of its 140,000 annual visitors arrive on cruise ships and never set foot on land. Once at Bartlett Cove, backcountry users have three ways to travel deeper into the park. The most obvious is to paddle north into Muir Inlet or the West Arm. But keep in mind that for the average kayaker it is a 3- to 4-day trip to reach the heart of Muir Inlet and an even longer paddle from Bartlett Cove to Tarr Inlet in the West Arm.

The most popular means of travel into the backcountry is on board one of the park's two concessionaire-owned tour boats, which run sightseeing trips into each arm. The *Thunder Bay* departs from the park dock in front of the lodge at 8 A.M. daily in the summer for an 8-hour trip up Muir Inlet. For the price of the round-trip ticket plus drop-off and pick-up fees, hikers and kayakers can use the vessel to reach designated points up bay, which traditionally have been at Muir Point, Wolf Cove, and Riggs Glacier. The vessel,

Park tour boat in the West Arm

with a shallow-draft bow, noses close to the shoreline and then discharges backpackers directly onto land by way of a long ladder.

The *Glacier Bay Explorer* also departs daily but spends each night in Tarr Inlet and returns the next morning. The boat drops off and picks up kayakers and campers at Lamplugh Point near the glacier of the same name and the mouth of Johns Hopkins Inlet and Reid Inlet. This puts paddlers in the heart of West Arm, regarded by most as the more dramatic of the bay's two main inlets. Keep in mind that exact location of drop-offs in both arms may be changed in the future by the National Park Service in an effort to avoid destruction of an area's landscape by overuse.

Kayakers can either use the tour boats for both drop-off and pick-up or reduce the cost by taking the vessel up bay and paddling back. For more information about the tour boats, contact Exploration Holiday Cruises, 1500 Metropolitan Park Building, Seattle, WA, 98101; call 800 426-0600 (toll-free) or 206 624-8551 in Seattle.

From Bartlett Cove it is also possible to charter a floatplane for transportation up bay—the quickest but most expensive means of travel. For rates on fly-ins, contact the park concessionaire at Glacier Bay Airways, Box 1, Gustavus, AK, 99826; call 907 697-2249.

Getting to and from the Outside Coast

Most backpackers hike the Cape Fairweather traverse from Lituya Bay north to Dry Bay to have the prevailing winds at their back. The high cost of getting to and from the outside coast, more than anything else, keeps this trek a remote one that is attempted by only a few people each summer. The least expensive way to reach this section of Glacier Bay is by a chartered plane from Gustavus northwest to Lituya Bay on the outside coast. A second charter from Yakutat should be arranged for a pick-up at Dry Bay, which is located south of the small Alaskan town. Once at Yakutat, you can get a commercial flight to Juneau or Seattle through Alaska Airlines. As mentioned earlier, the appendix lists air taxi operators for each area.

PADDLING TO GLACIER BAY

With your own kayak and plenty of time, you could spend a summer using the Alaska Marine Highway System to leapfrog from one wilderness paddle to another. Unfortunately (as stated earlier), the state ferry does not stop at either Bartlett Cove or Gustavus. Its closest port is Hoonah, situated across Icy Strait on the north shore of Chichagof Island. The least expensive way to get a hardshell kayak into the park is to purchase passage on the *Glacier Express,* which allows kayaks to be carried aboard.

Those contemplating paddling to the park from another designation in Southeast Alaska should first consider the risks involved. Although every summer a handful of kayakers make the journey, it is an open-water paddle for veteran ocean kayakers only. Such an adventure requires previous experience with rough water conditions and knowledge about the extreme tides and currents you will be encountering.

The easiest route for this hazardous journey is to begin at Auke Bay (north of Juneau) and paddle Saginaw Channel between Shelter Island and the north tip of Admiralty Island. At this point Lynn Canal averages 3 miles of open water between Mansfield Peninsula (on Admiralty Island) and the mainland.

Another possibility is to take the ferry to Hoonah and then paddle across Icy Strait from Point Adolphus to the mainland. However, it's an extremely tricky paddle through almost 7 miles of open water. Strong currents and tides, including standing waves, make it a challenging crossing. Some paddlers reduce the amount of open water by traveling from Point Adolphus northeast to Pleasant Island, a 4.5-mile trip. From there they cross the channel to Gustavus and follow the coast west around Point Gustavus into the park. There are often strong tidal rips near the point. Kayakers unsure of the situation will land on the beach east of the area and hike around the point to study the currents or wait for slack tide.

4 EXPERIENCING THE BACKCOUNTRY

The elements of Glacier Bay were painting pictures on a revolving stage. I was weathering out a downpour on the beach of Wolf Cove, with time on my hands and space for my eyes and mind to wander.

A decade ago I rounded Wolf Point and was greeted by a shoreline full of icebergs, all stranded until the next high tide took them farther south. Today only the largest bergs were making it past the point, and they were floating by, one at a time. The beach was bare.

Lunch was already consumed and it was much too early for dinner, so I huddled under the kitchen tarp and spent the afternoon listening to the rain roll off the canvas while watching this parade of icebergs float past me. Muir Inlet was a scenic stage for their procession. The clouds were low, hanging like a layer of cotton above the water and swaddling the mountain peaks that usually dominated the horizon. Instead of peaks, there were three layers to stare at—water, space, and clouds, one very evenly sitting on top of the other.

One by one the icebergs appeared from around the rocky point and floated past me. At first they were only blue ice. Then one definitely had the shape of a seal when it lifts its head to study an approaching kayaker, and the next was a camper hunching over in an effort to revive a campfire. After that they each took on an identity of something else: a ship, the fluke of a whale, a Canada goose, a small house.

I must have watched them for two hours. When they began to appear again as only icebergs, I stood up and wandered to my tent for an afternoon nap in the steady rain.

THE COASTAL KAYAKER

On rivers kayakers have whitewater, but throughout the coastal areas of Southeast Alaska the most challenging water is a different color. Locals call it bluewater—the tidal waters found in the bays, fiords, and straits that form the shoreline of coastal Alaska. Bluewater is not foaming rapids but is characterized by extreme tide fluctuations, cold temperatures, and the possibility of strong currents or tidal rips. In Glacier Bay, like in most of coastal Alaska, the open canoe gives way to the kayak, and bluewater paddling is the most frequent means of escape into this icy wilderness.

This is not whitewater kayaking with its helmets, wetsuits, and ultralight, streamlined boats. In bluewater, or ocean touring, the eskimo roll is not a prerequisite to any trip; in fact, the art of rolling an overturned boat to its upright position is beyond the ability of most paddlers in loaded doubles and folding boats. Ocean-touring kayaks, for the most part, have the distinction of being more stable, wider in the hull, and a better tracking boat than their whitewater counterparts. These advantages are needed by bluewater paddlers who find themselves carrying more gear, tackling wider stretches of water, and using the boat as a means of transportation into a

Kayak along the shore of Johns Hopkins Inlet

wilderness rather than a way of surviving the Class IV rapids of a constricted canyon.

In Glacier Bay, a kayak is the overwhelming choice over the traditional open canoe, which is rarely seen up bay. Kayaks are lighter to handle than most canoes; they track better and are less resistant to wind because they ride lower in the water. Add a spray cover or skirt that seals the paddler in the cockpit, and kayaks become watertight, an important feature during an all-day drizzle or a sudden rise of waves.

Yet veteran canoeists will find Glacier Bay an attractive place to undertake their first kayak journey. A concessionaire at Bartlett Cove rents kayaks while the park tour boats offer a drop-off and pick-up service for paddlers, providing backcountry users with easy travel in normally calm conditions up bay, especially in Muir Inlet. But both experienced bluewater paddlers and canoeists-turned-kayakers must keep in mind the special conditions and dangers found in Glacier Bay.

Needless to say, the water in Glacier Bay is very cold, usually ranging from 34 to 42 degrees Fahrenheit. This glacial seawater makes capsizing more than unpleasant; it is life-threatening. If renting a kayak, practice your strokes in Bartlett Cove before heading out on a tour boat, and discuss self-rescue methods with NPS rangers. Even if you have a personal flotation device, survival time in the water is less than two hours; without a life jacket it is considerably less. If you do flip, stay with the boat and attempt to right it and crawl back in. Attempting to swim to shore, even if you are only a couple hundred yards away, is foolish at best.

Most paddlers try to route their trip to follow the shoreline as much as possible, and then stay within a quarter-mile of it. Major crossings of over a mile should be done with a keen eye on the weather and in the calm conditions found in early morning or at dusk. Most important is to head in and "hole up" when wind and rough weather make the paddling too risky. With a spare day or two, paddlers rarely have problems meeting their designated pick-up spots and times. You should throw away the daily, hour-by-hour schedule you depend on so much in the city. Instead, flow with the wind and the tides, and stop when the weather becomes too foul or the scenery too irresistible to pass up.

Give a wide berth to marine animals that are sighted during a paddle—sea lions, seals, orcas, or other whales. The Endangered Species Act makes it illegal to be closer than a quarter-mile from a whale. Even if there weren't a law, however, caution would dictate staying a good distance from anything so large. Rarely do whales intentionally afflict harm on boaters, but an untimely encounter with one surfacing could easily result in a capsized kayak. Paddlers in areas of tidewater glaciers should be especially careful not to frighten female seals and pups on floating icebergs during pupping season in May and June, and should avoid all bergs with blood on them, as they are sites of recent births.

The enormous icebergs that drift slowly by should be treated with as much respect as the whales. Beautiful and intriguing from afar, they can be dangerous if approached too closely. Icebergs are notorious for turning and breaking without warning, and larger ones cause a rippling of strong waves. Paddlers should keep a safe distance from floating ice and never attempt to climb onto it.

Calving glaciers themselves are just as active, and the National Park Service recommends that kayakers stay at least a half-mile from their faces. It is not only the falling ice that paddlers should be aware of but, as with icebergs, the chain of waves that can be created as ice suddenly breaks off and falls into the water. The rolling swells will also affect anybody resting or camping on nearby shorelines. Kayaks that are not carried upshore or tied down to a rock can be lifted and swept out in seconds. The shoreline across from Margerie Glacier in Tarr Inlet is a classic example—it's a stunning place to pitch a tent but is frequently washed over by large calvings.

Icepacks, the gathering of bergs in front of tidewater glaciers, act as barriers that prevent paddlers and boaters from venturing any closer. The most extensive icepack is in Johns Hopkins Inlet; it is so thick that most paddlers are stopped 4 to 5 miles short of the glacier. Kayakers can weave their way through the start of most icepacks, but keep in mind that the constant bump and grind of small bergs will put cracks in hardshells and could totally destroy the front of a folding kayak such as a Klepper or Folboat. The other thing about icepacks is they tend to string out during low tide, creating many channels and passages that are inviting to paddlers. But once the incoming tides begin to flow, unsuspecting kayakers may find the clear corridor gone and be trapped for up to 6 hours in an impenetrable sea of crushed ice.

Unlike in canoes, framed backpacks are useless in kayaks and gear is

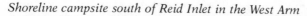

Shoreline campsite south of Reid Inlet in the West Arm

better stowed in duffel bags or small day packs. If you don't already have a set of waterproof storage bags, bring a large assortment of plastic bags, including zipper-lock bags and large garbage bags. All gear, especially sleeping bags and clothing, should be sealed in plastic bags and then placed in a duffel bag or stuff sack. Water tends to seep in even when the weather is calm and your spray skirt taut.

TIDES

Besides topographical maps and a compass, all paddlers should have a Southeast Alaska tide book, the bible among kayakers in Glacier Bay. Maps and tide books are both available from the National Park Service at Bartlett Cove. Tides can range as much as 25 feet in the bay, and in many narrows swift currents and whirlpools are created during mid-tides. Sitakaday Narrows in the lower bay is famous for its rips, but it is only one of the many places kayakers should enter at slack tide.

Kayakers should use the Juneau tide schedules, but for complete accuracy you need to turn to the tidal differences and then add time to determine the high and low tides for the various points listed. Within Glacier Bay there are four areas to consider; paddlers should select the closest one and then add the additional minutes to the Juneau tide to determine the exact time of the local tidal influx:

PLACE	TIME		FEET
	HIGH	LOW	
Bartlett Cove	+0:11	+0:12	−1.6
Willoughby Island	+0.24	+0:36	−0.2
Muir Inlet	+0:29	+0:39	−0:3
Composite Island	+0:28	+0:37	+0.3

Life is easier when you schedule your paddling around the tides and use the current to help you along. Even if it means beginning at the crack of dawn or waiting until 1 P.M., it is well worth waiting for the right tide. The days are long in Glacier Bay but the tides last only 6 hours.

When going ashore or quitting for the day, always carry your boat above the highest seaweed line and tie the bowline to a heavy stone or log. Nothing is more frightening than watching your kayak float out to sea with the tide.

HIKING IN A TRAILLESS PARK

Outside of Bartlett Cove, there are no developed trails in Glacier Bay. But there are beaches, moraines, river banks, ridges, and glacier remnants, all offering natural routes away from the inlets and into the park's interior. The challenge to hiking in Glacier Bay is the park's ever-changing state of succession.

The greatest challenge, as many hikers see it, is alder; tangling, frustrating, kick-you-in-the-butt, slap-you-on-the-head, alder. What was a sterile ravine a decade ago may now be a mass of alder so thick and heavy that it forms an impenetrable barrier for most backpackers. Bucking alder on a cross-country trek is inevitable in Glacier Bay. But consider carefully what you are willing to endure before setting out to climb through a ridge of it. Anything over a half-mile is a lot of alder to battle, and something that shouldn't be attempted on a spur-of-the-moment hike in the evening.

The other thing to remember is that where there's alder, odds are there's also devil's club. This shrub, which can easily exceed 8 to 10 feet in height, has large oaklike leaves whose underside and stems are covered with sharp spines. Painfully and difficult to remove once embedded in the skin, the spines cause hikers to steer clear of the plant or else wear their heaviest longsleeve shirts and pants when crashing through it.

The alder, and to a lesser degree willow and devil's club, can form a lush jungle that prevents many areas from being enjoyed. But hikers find the banks of glacial streams and rivers open avenues into the backcountry. Many will have shores of soft glacial silt that makes for easy walking while others will be lined with huge boulders where hikers will need sturdy boots to hop from one to the other. Be prepared when following a river to scramble up a few high banks or cliffs where the waterway bends sharply toward your shore.

But eventually in your backcountry treks you will have to cross a meltwater stream. Such streams and rivers are usually turbulent and so silty that you cannot see the bottom to judge their depth. It is best to cross them in old tennis shoes with a sturdy shaft in hand while unhooking your pack's hip belt for quick escape in case of a spill. Select a wide section, which is usually the shallower part of the river, and cross diagonally downstream. Rivers that quickly reach your thighs and are splashing at your waist should be considered carefully before continuing the crossing. If you know in advance of a major fording, plan to cross in the morning when water levels may be lower (sometimes by as much as a foot) rather than in the evenings. Rivers will be at their lowest levels during May and June, after which they continue to rise as the snow melts throughout the summer.

In several areas of the park, most notably the Wolf Cove region, backpackers rejoice in the unique hiking and strange world that glacial remnants offer. Ice remnants are often called dead or dying glaciers because they have become isolated from the main trunk or simply have become stable, with little activity. The surface itself is rough ice that is extremely abrasive and demands sturdy footgear for crossing. On misty or rainy days a glaze often forms on the ice; on a cloudless day hikers find it necessary to pack along dark sunglasses, or "glacier goggles" as many call them, to endure the strong glare.

Hiking on remnants and black ice (old glacial ice that has been covered with gravel and stones) is relatively safe, but does warrant some caution. Hikers should travel in pairs and never walk across glacial ice still covered by last winter's snow. Lingering patches should also be avoided, as snow has the ability to hide crevasses (great cracks in the glaciers) and moulins (melt holes that drop deep into the ice). A muffled roar of water is a sign that a meltwater stream is nearby; often the ice covering it is quite thin. Crampons for the soles of boots and ice axes are not usually needed when hiking the remnants, but having a good length of rope is a reassuring precaution in case an unexpected mishap occurs.

The most difficult part about hiking on remnants may be the act of ap-

Hiking outwash up to Scidmore Glacier

proaching them, since the edge of the old glacier is often thin and undercut by streams. The sediment in front of the ice is so saturated from meltwater that often it becomes "quick mud," which at times has been known to swallow hikers up to their knees. Use a staff to check out such questionable goop, and once you encounter it, move quickly to avoid sinking.

It is also best to simply view rather than enter the ice caves found near the edge of glaciers. Dangers lie in the rocks embedded in the ceiling, in ice slaps and pinnacles that could suddenly give away, and in meltwater streams that could break through.

Traveling on active glaciers requires advanced experience, proper equipment, and expertise in ice climbing. Most backpackers arrive at the park without such preparation, in which case they should by all means avoid these active rivers of ice.

Those who plan to explore the outside coast of Glacier Bay will encounter yet another set of hiking conditions. The Cape Fairweather traverse involves hiking long stretches of sandy shoreline, rounding points and capes composed entirely of boulders, and using bear trails that dip away from the ocean and back into the lush forest. Sturdy boots with good ankle support are necessary, as hikers will find themselves "bouldering" often—in some areas the rocks will be 4 to 5 feet tall. The constant leaping from one boulder to the next is hard on the ankles and calves of any hiker, but especially those with poor footwear. In some places along the Outer Coast, however, you might just as well be walking down miles of sandy beach in bare feet.

Every group attempting the route should have two things: a two-person raft with paddles, foot pump, and repair kit; and 400 feet of rope. The raft will be used to cross several rivers and streams that are too deep and fast to ford. The rope combined with the raft can be used to set up a ferry system to carry people and equipment safely across the extremely turbulent glacial streams that may be too swift to paddle. Rivers on the outside coast are constantly changing their courses, water volumes, and velocities. Hiking parties must be prepared to turn back in the event of encountering one too swift for even a raft.

Drying out in Scidmore Bay after three days of continuous rain

MAKING CAMP IN THE WILDERNESS

Before departing up bay, check with NPS rangers for the areas in the park that are closed to foot traffic and camping. Traditionally Marble Islands, Sealer Island, Lone Island, the two islets south of Russell Island, and many treeless islands near Bartlett Cove have been designated as off-limits to protect resources and the summer nesting sites of various species of birds. Sandy and Spokane coves are also closed to campers from May 15 to July 1, owing to heavy concentrations of black bears during that period. A permit is required to camp in the backcountry, and all closures will be pointed out when you are issued one.

There are many other things to keep in mind before you pitch your tent. Many parts of the park do not offer good campsites: either the alder or other vegetation is too thick, or the landscape, particularly in the inlets of tidewater glaciers, is too raw and steep to enable you to find enough level space for even a small dome tent. When hauling your equipment and boat up to the first dry spot, you will find that coves bordered by vast tidal flats are a difficult trek at low tide.

The best places tend to be the naturally disturbed areas such as outwash fans, beach zones, or the banks of a glacial stream. The National Park Service urges all backcountry users to camp out of sight of water and especially to avoid camping on beaches or coves along which animals have established major thoroughfares. Make sure you pitch your tent well above the high-tide line and not just at the last strip of seaweed on the beach; even the ryegrass zone becomes flooded during the highest tides. Waking up at 2 A.M. in a puddle of seawater is a shocking way to learn how far the tides can rise.

Finding drinking water is rarely a problem except when camping on islands or high ridges. If you plan to paddle into the Beardslee Islands, take a good supply with you from Bartlett Cove and then be resigned to paddling to the mainland to restock later on. Rainwater is another source of drinking water but an unreliable one because of Glacier Bay's unpredictable weather.

Although nothing could seem purer than the meltwater tumbling from a glacier, the National Park Service considers *Giardia lamblia* (an intestinal parasite) to be present in some areas and therefore recommends that backcountry users treat all water. This is possible by either boiling the water for one full minute or running it through one of several filters on the market today designed to remove the parasite. The water from glacial streams that are gray with silt is drinkable, but backpackers would do well to let it stand overnight to let the till settle to the bottom. For some people, consuming water with this glacial silt still in suspension has the same effect as a good laxative.

All garbage should be hauled out of the backcountry, but human waste poses special problems. It does not decompose effectively in the thin organic soils and cool temperatures of Glacier Bay, and many popular camping spots, such as Riggs Glacier, are not big enough to adequately distribute the waste of all who stay there.

Unorthodox as it may seem, the best place to recycle human waste is below the high tide line along the wide open shore or outwash fans of glacial streams. There may not be much cover, but here the currents can carry it away into the tidewater where marine organisms break it down to basic

Shoreline campsite south of Reid Inlet in the West Arm

nutrients. Toilet paper, however, should be burned, and anglers should drop their fish scraps into deep water. Many campers find the tidal areas ideal spots to cook and eat, since all traces of food and crumbs, which might attract bears, will be flushed clean every six hours.

When departing a camping area, not only pack up all garbage, but make it a practice to "naturalize" the area by replacing rocks and scattering forest litter back where you found it. Allow other campers the joy of pitching their tent in a pristine area free from man's heavy footprint.

BEARS

The brown (grizzly) bear, with its trademark hump and massive size, instills the most fear in people, but the black bear is usually the biggest nuisance. Both bears are present in Glacier Bay: The brown bear being most common on the park's outside coast and in the upper reaches of the West Arm; the much smaller black bear is present almost everywhere else. Keep in mind there are also large black bears and smaller, sub-adult brown bears, and occasionally a member of one species may be confused for the other.

An encounter with a bear is a memorable experience for most visitors. Still it is best when in the backcountry to take precautions to avoid close encounters with them. When hiking in heavy vegetation where visibility is limited, prevent a surprise meeting by whistling, singing, or talking loudly. And in areas where you notice their presence—tracks, bear trails, unearthed roots, or scat—seek out open country for travel. Most important, stay away from cubs, as a sow will attack to protect them.

When camping, make sure you don't pitch your tent on a bear trail or in the middle of the animal's favorite berry patch. Bear trails are easy to identify; some seem permanent, as the animal occasionally walks the same path right down to the exact steps, leaving well-defined marks. Camp in one spot, cook and store your food somewhere else—at least 100 yards away. When retiring for the evening, hang your food in a tree, at least 10 feet above the ground and 4 feet away from the trunk, when possible. If you are up bay, where trees are few, either store the food on a rocky outcrop off a ledge or cover it with a rock pile. Wash your hands and face before entering the tent, and take no food, cooking utensils, clothes you wore while cooking, or odorous substances (such as toothpaste or hand lotion) in with you.

If you encounter a bear, make a wide detour around it and cross upwind, allowing its nose, the bear's most powerful sense, to detect your presence. A bear on its hind legs is only trying to perceive you better, not preparing for a charge, as many people believe. If a bear does move in on you, maintain a calm, assured posture and in an authoritative voice, talk to it. Most charges, experts say, are often just bluffs, ending abruptly with the animal veering off in one direction. Don't panic by turning around and fleeing; you'll never outrun a bear. A last resort is to lie face down in a fetal position (protecting your neck with your hands and arms) and remain silent, as difficult as this may seem. Experts say that bears only want to exert their dominance. Many charges have ended with the animal merely sniffing (and departing) or inflicting only minor injuries.

BACKCOUNTRY NEEDS

What you should bring in the way of gear depends, of course, on what you plan to do. Visitors who simply want to camp near the face of Riggs Glacier for a few days need far less than do backpackers planning to complete the Cape Fairweather traverse. Whether you plan only to spend one night in the backcountry or paddle for three weeks in the remote West Arm, you have to be totally self-sufficient and prepared to handle any mishap that might occur.

Just as important as deciding what to take is organizing your gear carefully before you leave home. Attention to small details then prevents you from arriving in Bartlett Cove either overburdened with too much gear or, worse yet, missing an essential piece of gear such as a camp stove or rain gear. Neither the park headquarters nor the general store in Gustavus sells much in the way of camping equipment and supplies. About the only things you can count on purchasing at Bartlett Cove are topographical maps and white gas for stoves (the airlines will not allow the latter on their planes).

FOOTWEAR: Those restricting their trip to hiking and camping up bay at Wolf Point or Riggs Glacier will want sturdy hiking boots for trekking across rugged ridges or glacial remnants and a pair of old tennis shoes for fording rivers. Kayakers will want 14-inch rubber boots, preferably lined with wool insoles, for the constant hopping in and out of their craft. Paddlers, who will be spending most of their time on the water, will find the traditional leather hiking boot too bulky and heavy for the occasional hike they take. Instead, many kayakers prefer the newer lightweight nylon boots.

TENT: Free-standing dome tents are the type most frequently seen in the backcountry, and they work well up bay where the terrain of rocks and boulders consumes stakes daily. But any backpacking tent will do as long as it has two things: a rain fly for the ceiling and insect netting on the doors and windows. In addition, people who plan to be out for weeks at a time might consider bringing a rain tarp for their kitchen and for keeping the rest of their gear dry.

RAIN GEAR: Glacier Bay averages between 75 and 125 inches of precipitation annually, so good foul-weather gear is crucial. That means a rain jacket and pants, a rain fly for the tent, a pack cover for extensive hikes, and a spray skirt for kayakers. One-way permeable garments are ideal for the many days of mist and light rain, but they shouldn't be your sole protection against the downpours you will eventually have to endure. In heavy rain, these garments have a tendency to flood through and leave you soaked and chilled on the inside. By the same token, heavy-duty rain gear with poor ventilation will cause overheating and excessive perspiring. Ponchos, which are awkward to hike and paddle in, are not the answer. The best solution seems to be gear that has some ventilation under the arms, and a wool shirt underneath that will absorb excess moisture and still keep you warm.

CLOTHING: Prepare for cool, rainy days and then rejoice when the sun breaks out and the temperature soars into the high seventies. The predominant choice of material in the backcountry is wool or a synthetic equivalent, such as polypropylene or bunting, all of which can keep you warm even when they are wet. You almost can't bring enough wool to Glacier Bay. Every hiker or paddler should pack along a wool shirt, socks, mittens, pullover hat, and even a pair of lightweight wool pants. Select garments that will allow you to dress in layers. A lightweight shirt along with a sweater and windproof parka are much more comfortable during a long day of paddling than a heavy jacket.

SLEEPING BAG: A medium-weight sleeping bag should be sufficient for Glacier Bay during the summer. The bag should have some kind of synthetic filling such as a polyester fiber because even when wet these bags provide insulation. If need be, they can be wrung half-dry. Goose down should be avoided because it clumps when wet and loses most of its insulative ability making for a long and cold night. Backcountry users should also bring along a sleeping pad or foam-rubber mat for added comfort against the often hard and rocky ground.

CAMP STOVE: No matter where you go in the park, a camp stove is a necessity. In the lower regions, where there are trees and down wood, the rainy weather can make fire building a long and frustrating task. Up bay near the glaciers in Muir Inlet and the West Arm, there simply won't be any wood available except for an occasional piece of driftwood or interglacial wood, which campers are prohibited to burn.

FOOD: Backpackers should bring all the food they will need during their stay in Glacier Bay, as there is no store in Bartlett Cove and only limited supplies available at a general store in Gustavus.

INSECT PROTECTION: This is Alaska and that means there will be bugs to contend with. Although the problem is not as serious as in Alaska's interior (you won't see people draping mosquito netting from their hats), you still need protection. Gnats and flies tend to be worse than mosquitoes, but you'll also run into no-see-ums, white sox, and deerflies. Bugs begin appearing in late June, peak in mid-July, and taper off in mid-August. Rarely will you be bothered by them while paddling, and beaches usually have enough breeze to keep them from swarming around you. But bucking your way through a jungle of alder in the middle of July is another story. Bring repellent and double-check all your screening in your tent. You can handle bugs during the day as long as you don't have to contend with them in your tent at night.

MAP AND COMPASS: Every group paddling and hiking in the backcountry should have the proper topographical maps (or nautical charts) and a compass, along with the knowledge of how to use the two together. This is not a place where you follow a trail; getting turned around is easy in Glacier Bay.

PADDLING NEEDS: An extra paddle and lines for the boat are not only good insurance, but also handy when you want to prop up your rain tarp and there's no brush around. Kayakers should also have sunglasses and sunscreen for the intense reflection off the water on clear days; and a bailing device, whether it be a hand pump or just a sponge, for each boat. The National Park Service stresses that every paddler should have a personal flotation vest. Within each kayak party there should be a small kit consisting of duct tape, a fiberglass repair kit, or silicone rubber cement for fixing unexpected cracks in a boat.

WEATHER AND WHEN TO GO

The park is open year-round, but the visitor season runs from mid-May until mid-September, coinciding with when accommodations and concessionaire services are available at Bartlett Cove. The flow of hikers, campers, and paddlers into the backcountry reaches a peak from July through early August. Reservations for kayak rentals or spots on the tour boats are needed throughout the summer.

Whenever you arrive, prepare for the cool, damp weather of a maritime climate and accept the fact that a bright, sunny day is a rare treat. As a rule, the least amount of rain is usually encountered in May and June; then the precipitation steadily increases, leading to a rainy late August and September. But that doesn't mean you won't be weathered-in by three days of hard showers in June, enjoy an 80-degree, cloudless afternoon in August, or even endure a snow fall in late May.

Come prepared for anything and look for the beauty in the gray cloudy days that are typical in Glacier Bay. Here the park seems to brood in shades of grays and blues as the clouds hang low above the water, forming perfect layers, and the mist comes and goes intermittently with the wind.

Timing your trip depends on personal taste and what you intend to do. In early June you have few crowds, even fewer bugs, long days, and glacial streams that are low, making them easy to ford. But snow persists on the ridges, mountain passes, and ice remnants, making them inaccessible to most hikers. By August the highlands are free of snow, but what was a melt-water stream in June is now a torrential river. The National Park Service has characterized the visitor season as follows:

> *Spring and Early Summer:* Snow persists, especially in the high country, and ice fills the inlets in the upper part of the bay. Mountain goats are down low, birds are returning north, and the plants are beginning to turn green again.

> *Mid-Summer:* Whales are gorging on feed in the bay, and the birds are raising their fledglings. The sea birds can be seen in variety and abundance. Wildflowers in vast array color the landscape. It is also time for insects to prosper.

> *Late Summer to Early Fall:* Snow is mostly all gone from the ridges and low peaks, but the rain and winds increase. Berries are abundant and the salmon migrate into the rivers. Vegetation shows off its splendor of fall colors. Whales and summer birds begin to leave. The aurora borealis appears during starry nights.

PERMITS, SERVICES, AND INFORMATION

All paddlers, hikers, and campers should obtain a backcountry permit once they arrive at the park and before they depart into the backcountry; only one permit is needed for each party. Obtaining this permit means that the National Park Service has a record of your itinerary; it also helps facilitate the exchange of information on conditions and hazards in the backcountry. The registration should be done at the Visitor Information Station at the foot of the Bartlett Cove dock. The station is also the place to purchase topographical maps, nautical charts, books, and literature about the park.

Although the National Park Service manages Glacier Bay, it contracts out most of the visitors' services available in Bartlett Cove to concessionaires, including hotel accommodations, tour boats that cruise up bay (see chapter 3), and kayak rentals. The park service does maintain a walk-in campground that features a bear-proof cache, fire pit, firewood, and an eating shelter in a scenic setting along the shore, a quarter-mile south of the dock. The campground is free and no reservations are needed; stays are limited to 14 days maximum.

Park naturalists organize a variety of daily programs, including hikes into the lush forests surrounding Bartlett Cove. They also present nature films and slide-illustrated talks in the evening at the park lodge. A naturalist greets all campers arriving at Bartlett Cove by bus and conducts a camper orientation program that provides information on where to go in the backcountry, minimum-impact camping, and safety precautions.

LODGE ACCOMMODATIONS: The 55-unit lodge offers picturesque and comfortable rooms along with less expensive hostel-type accommodations that are charged on a per-bed basis. Also available to the public are a restaurant, coin-operated laundry and showers, fuel, a gift shop, and a lobby with a fireplace. The Glacier Bay Lodge operates from mid-May until mid-September and is the only hostelry in Bartlett Cove; other accommodations exist in Gustavus, 10 miles from the park. The lodge is booked heavily during the summer, and persons interested in a room are advised to write well in advance for reservations. For information and reservations, write or call Glacier Bay Lodge, 1500 Metropolitan Park Building, Seattle, WA, 98101. Call 800 426-0600 (toll-free) or 206 624-8551 in Seattle.

KAYAK RENTALS: Available for rent in Bartlett Cove are two-person, hardshell kayaks with foot rudder systems, skirts, personal flotation vests, and paddles. You can rent them for a day to paddle around Bartlett Cove, for short trips into the Beardslee Islands, or for long paddles into Muir Inlet or the West Arm. They can also be rented and loaded on board the tour boats to be dropped off up bay, thus eliminating the long haul from park service headquarters to the first tidewater glacier 45 miles away.

During the summer the kayaks are picked up daily at 9:30 A.M. and 6 P.M. at the Visitor Information Station at the head of Bartlett Cove dock, at which time an orientation meeting for kayakers takes place. The number of boats for rent at Glacier Bay is limited; advance reservations are strongly recommended, especially from early July through mid-August. For more information or reservations, contact Glacier Bay Sea Kayaks, P.O. Box 26, Gustavus, AK, 99826; call 907 697-2257.

MORE INFORMATION: While planning your trip, you can obtain information about Glacier Bay or about kayaking up bay, hiking the outside coast, or running the Alsek River, by contacting the park headquarters at Glacier Bay National Park and Preserve, Gustavus, AK, 99826; call 907 697-2230.

Juneau, the capital of Alaska, is the major departure point for the majority of travelers heading toward Glacier Bay. While in town, go to the Centennial Hall Information Center, 101 Egan Drive (a half-mile from downtown) for more information about the park or any monument or preserve in Southeast Alaska. During the summer the center is open daily from 8:30 A.M. to 6 P.M. Call 907 586-8751.

For more travel information, whether it be for airlines, lodging, state ferry transportation, or other sights in Alaska, contact the Alaska Division of Tourism and request its annual *Alaska Travel Planner* guide to the state. Write to Alaska Division of Tourism, Pouch E-600, Juneau, AK, 99811.

Books and other literature that provide further information on the park's glaciers, geology, wildlife, and natural history are available through the Alaska Natural History Association, a nonprofit organization that works with land-managing agencies in Alaska. Contact Alaska Natural History Association, Glacier Bay National Park and Preserve, Gustavus, AK, 99826; call 907 697-2230.

PART II

GLACIER BAY BY PADDLE

5 EAST ARM

In the early morning mist of Muir Inlet, a lone kayaker stopped paddling when he sighted the blow of a humpback whale. He heard it first, then spotted the explosion of air a mile north of him. He was awed even more when the animal breached, heaving its massive 40-ton body out of the water in a spectacular display of acrobatic leaps and turns.

Then it stopped. The whale could have departed in any direction, but moving toward the kayaker was a line of air bubbles and a smooth patch in the rippled surface of the water. He put down his paddle; he knew where the whale was heading, and he knew he was required by law to maintain a 0.25-mile distance. But the watery trail continued to move in his direction, so he picked up his camera.

Suddenly his viewfinder went black; the whale had surfaced 20 yards to his left. His heart was pounding, his adrenaline flowing and his shutter finger instinctively went to work. But in the excitement of being so close, he fumbled momentarily with the winder. In one second, he cursed the lack of a power winder, in the next he watched the photo of a lifetime pass before him: the whale's enormous fluke high in the air, shaking in a rainbow of mist and water, giving one last powerful thrust that would send the animal to the depths of the bay.

A photo lost but a vivid memory etched in his mind. It was a good 15 minutes before he picked up his paddle again.

Riggs Glacier in Muir Inlet

EXPLORING THE EAST ARM

The destination for many kayakers is upper Muir Inlet or the East Arm. With a drop-off and pick-up by one of the local tour boats, *Thunder Bay* (see chapter 3), they can spend three or four days exploring the upper portions of the narrow inlet and the barren terrain around Riggs, McBride, and Muir glaciers, all spectacular tidewater glaciers. The exact drop-off points in Muir Inlet will change over time, as the NPS plans to gradually rotate them to allow for site restoration. All kayakers should contact the NPS for current drop-off sites and then plan their trip accordingly.

Long-time, traditional drop-offs, which could be altered in the future, include areas near Wolf Cove, Riggs, Glacier, and Muir Point at the entrance of Adams Inlet and south of Wachusett Inlet, two inlets offering a wilderness solitude that is sometimes hard to find around Riggs Glacier. To paddle in and enjoy either inlet requires at least two days, preferably three.

To paddle up bay, plan on three to four days to move steadily from Bartlett Cove to McBride Glacier, the first tidewater glacier en route. Although many kayakers utilize the tour boat drop-off service because of limited days, the paddle along the east coast into Muir Inlet is well worth the extra time as it adds another dimension to the trip. Here the story of Glacier Bay's succession unfolds day by day, almost stroke by stroke, making the long awaited encounter of the glaciers that much more dramatic and rewarding.

Bartlett Cove and Beardslee Islands

Park Headquarters to Beartrack Cove
Distance: 9–11 miles
Paddling time: 5–8 hours

Most kayakers depart from Bartlett Cove eager to reach the glaciated inlets of the park, and en route to them, either by tour boat or paddle, miss the charm and solitude that lie in the Beardslee Islands. This special group of islands, which begins with Lester Island, 1 mile north of the NPS dock, contains no glaciers or even a view of them. But they offer solitude and placid paddling in a scenic setting that makes for an ideal two- to four-day trip out of the park headquarters.

The paddling is easy, but go prepared. There is no water on the islands except for occasional surface runoff. Pack water with you and then search the mainland streams for additional sources. The islands have their share of black bears, so maintain a clean camp and put your food in a tree away from the tent. Go ashore and camp only on forested islands to avoid disturbing bird colonies on treeless ones, and above all, consult a tide book to time your departure from Bartlett Cove.

Kayakers enter the Beardslee Islands from the back door, paddling a channel that leads northeast from the head of Bartlett Cove. The narrow channel is formed by Lester Island to the west and the mainland to the east. At one time the channel was navigable at low tide but now it's not, a result of the rising land that affects much of the Beardslee Islands' topography. The best time to leave is two hours before high tide, to make use of the incoming flood.

From the NPS dock you pass Lagoon Island and a second NPS dock behind it, and you reach the channel in less than 1 mile. The channel is narrow at first but opens where the Bartlett River empties into it from the east. The

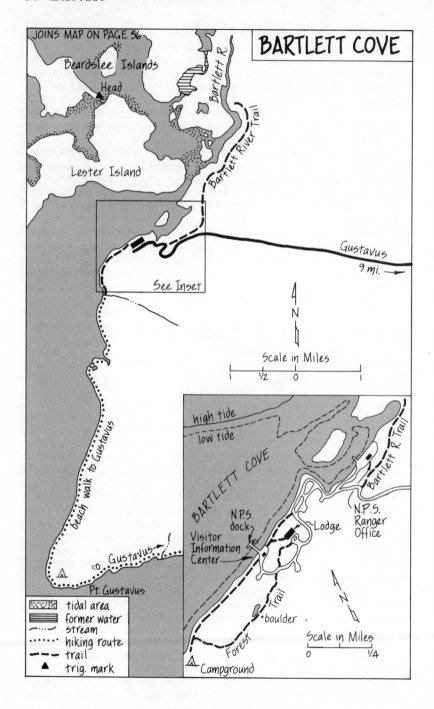

JOINS MAP ON PAGE 56

BARTLETT COVE

Beardslee Islands

Head

Bartlett R.

Bartlett River Trail

Lester Island

See Inset

Gustavus
9 mi.

N

Scale in Miles

1   ½   0   1

high tide

low tide

BARTLETT COVE

Bartlett R. Trail

N.P.S.
dock

N.P.S.
Ranger
Office

Lodge

Visitor
Information
Center

to Gustavus

Forest Trail

boulder

N

Scale in Miles

0   ¼

beach walk to Gustavus

to Gustavus

Pt. Gustavus

Campground

tidal area
former water
stream
hiking route
trail
trig. mark

mouth of the river is hard to distinguish as there are several islands blocking it. The channel to the Beardslee Islands curves to the northwest from here and ends at an island, unnamed on the topographicals. Before reaching the island, you pass what appears on the maps as a long, thin tidal passage into one end of Hutchins Bay, but it is no longer open because of the rising land.

Secret Bay

Secret Bay, which measures 1.5 miles from north to south, is formed by a ring of three large islands, with Lester Island to the east, Young Island to the west, and another unnamed one to the north. On the topographicals there appears to be three channels into the bay, but the northeast one (trig mark "Head") now requires a short portage even at high tide. One open entry point is the southwest channel, but paddlers should reach it at high tide and during good weather as the south shore of Lester Island can get choppy during strong southwest winds. The other entry point is between Young Island and the unnamed island to the north, which should be entered from the east and north rather than through the Sitakaday Narrows. These narrows are a waterway of strong tidal currents, especially at ebb tide when the tidal rips occur.

The bay itself is usually a calm, well-protected body of water and a haven for dungeness crab, and Lester Island features many fine sandy beaches. The spot makes a good overnight trip from Bartlett Cove, as it is a 5-mile paddle from the park headquarters through the southwest channel into the heart of Secret Bay. Kayakers should plan on another 5 to 6 miles of paddling and a short portage if they want to return by way of the northeast channel. Such a return involves reaching the narrows into Bartlett Cove near high tide.

Hutchins Bay

The small, unnamed island at the end of the narrows from Bartlett Cove marks two channels, with one heading northwest toward the open water of Beardslee entrance (a passage of whirlpools and rips during mid-tide) and the scattering of islands north of it. The channel to the north leads 2.5 miles into Hutchins Bay, a scenic body of water, especially its northern portion. At slack tide on a clear day the northern half of the bay is stunning, with the calm water reflecting the distinctive peaks north of Beartrack Cove.

Hutchins Bay ranges almost 6 miles from the closed inlet to the south to several narrow coves that extend into the mainland to the north. The western side is formed by Link and Kidney islands, the destination of many kayakers looking for a good campsite their first night out. The small island of Kidney with its sandy strip is one spot, while the south end of Link, a narrow knob with a few trees that overlooks the channel formed by Kidney to the south, also makes for a pleasant spot. Both are waterless, but several streams empty into Hutchins Bay from the mainland, with the major one located 1.5 miles east of the north end of Kidney Island. There is also a small runoff stream, though at times it may be dry, that drains into the northeast cove of Kidney.

Topographicals show a waterway into Beartrack Cove from the north-

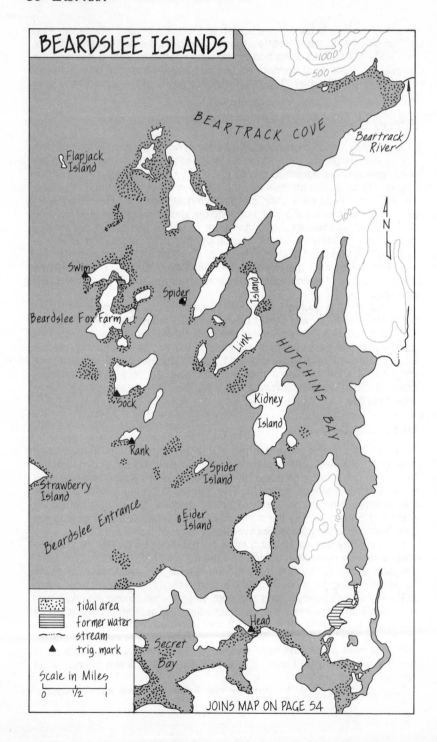

BEARDSLEE ISLANDS

BEARTRACK COVE

Beartrack River

1000
500
100

Flapjack Island

Swim ▲

Spider ▲

Beardslee Fox Farm

Link Island

HUTCHINS BAY

Sock ▲

Kidney Island

Rank ▲

Spider Island

Strawberry Island

Beardslee Entrance

Eider Island

100

N

Head ▲

Secret Bay

JOINS MAP ON PAGE 54

tidal area
former water
stream
▲ trig. mark

Scale in Miles
0   ½   1

west corner of Hutchins Bay, but this has long since dried up owing to uplifting land. Today the gravel strip makes for a 100-yard portage at high tide and a protected route into the major cove, as opposed to a paddle around the outstretched peninsula that forms Beartrack's south shore.

Western Beardslee Islands

To the west of Link and Kidney are the rest of the Beardslee Islands, stretching from Eider Island in the south to Flapjack, an island that is little more than a nice shoreline and a dozen trees off by itself 6 miles to the northwest. To protect bird colonies, the park service urges paddlers to avoid walking on Eider, Flapjack, and other treeless islands in the area.

In between Eider and Flapjack, the islands range in size and shape and offer kayakers countless stretches of shoreline and beaches to explore and camp on. Those shorelines that face the west provide a bonus on a clear night in late summer, as a sunset over the Fairweather Range can be a glorious end to a good day of paddling.

The only other Beardslee Island with a name is Spider, situated three-quarters of a mile southwest of Kidney. A mile west of Spider is an island (trig mark "Rank" on the topographicals) and a small point on its west side. The point is a splendid camping spot, with views of the Fairweather Range in one direction and the peaks north of Beartrack Cove in the other. While paddling to this island, keep an eye on the rocky islets and reefs that lie between it and Spider, and swing wide of any that are being used by seals.

To the north is a large island (the topographicals show it with the trig mark "Sock" on its south side) with good camping sites on its north shore. North of it are five more islands; the four to the west have been merged as the land has risen. Paddle to the east side of the first one (a round "island" on the topographicals), then look north for three old pilings in a small cove.

Oystercatcher on Geikie Inlet shoreline

This is the most conspicuous part of the former Beardslee Fox Farm. What is left is buried among the heavy bush inland. Camping is poor in this cove, but on the north shore (near the trig mark "Swim") there are spots with stunning views of the Fairweather Range.

To the east of the Beardslee Fox Farm lies a small island with the trig mark "Spider" (do not confuse with Spider Island to the south). Good camping lies a half-mile north off the treeless island on a sandy cove. From here you can follow the L-shaped peninsula north for 2 miles and pass through the gap formed to the west by a pair of islands, to swing into Beartrack Cove. Be aware that this channel is dry at low tide. Also be conscious that Beartrack can get extremely rough when a strong northwesterly wind is whipping down the bay, making the 2-mile open paddle across its entrance a rough one.

Heading Up Bay
Beartrack Cove to Muir Point
Distance: 20–22 miles
Paddling time: 2 days

From the tip of the L-shaped peninsula, Beartrack Cove swings 3.5 miles to the east, where the tidewater of the bay meets the mouth of the Beartrack River in an extensive tidal-flat area of soft mud. The northwest angle of Beartrack's mouth makes it a choppy place during northerly winds, which often spring up in early afternoon on clear days. Many campers avoid the extensive flats at the head of the bay and the resulting long walk to their boat at low tide by pitching their tents at a number of good spots along the shingle beach on the south shore. You may have to search for water, but it's worth it—an evening here is spent being dwarfed by the pair of 4,000-foot peaks that loom just to the north.

The coastline north of Beartrack Cove becomes a steep, wooded slope. There are few places to land and camp until you round the head (trig mark "Goat") and dip into the cove formed by York Creek, 3.5 miles away. The spot provides good campsites alongside a thunderous creek that roars out of the mountains into the bay. At low tide this cover turns into a wide gravel beach with pools of interesting marine life among the kelp and seaweed.

The shoreline remains steep beyond York Creek, with few places to land during high tide. As you head up bay, the Leland Islands can be seen to the northwest, while due north is the curved arm that separates Spokane and Sandy coves. The coves are a 5-mile paddle from York Creek and a break in the steep shoreline, which features an impressive set of cliffs right before you dip into Spokane Cove. Spokane is a pleasant place with a pair of streams emptying into it, but again, extensive tidal flats make camping unfeasible for most paddlers.

Good camping spots exist on the south shore of the peninsula that separates the two coves, and in Sandy Cove as well. Kayakers should be aware of the black-bear situation, however. The animals may congregate in heavy numbers from Beartrack Cove to Sandy Cove, especially in spring and early summer, when they are searching for their first meal after a long winter's nap. The Sandy Cove area may be closed to camping from mid-May to mid-July owing to a high density of aggressive bears; kayakers should check out the area's current status before leaving Bartlett Cove.

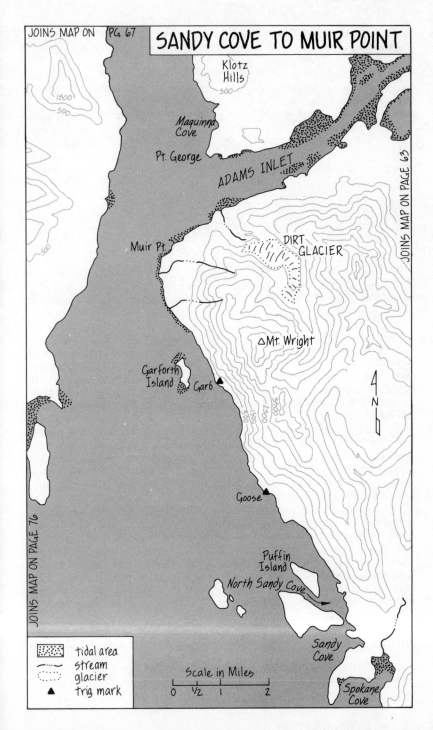

SANDY COVE TO MUIR POINT

JOINS MAP ON PG. 67

Klotz Hills

1500
500

Maquinna Cove

Pt. George

ADAMS INLET

JOINS MAP ON PAGE 63

Muir Pt.

DIRT GLACIER

△Mt. Wright

Garforth Island

Garb ▲

N

Goose ▲

Puffin Island

North Sandy Cove

Sandy Cove

Spokane Cove

JOINS MAP ON PAGE 76

tidal area
stream
glacier
▲ trig mark

Scale in Miles
0   ½   1   2

Seals in Johns Hopkins Inlet

A large island forms the north shore of Sandy Cove and from mid- to high-tide you can still paddle through the narrow gap it forms with the mainland toward Puffin Island and what is listed on some maps as North Sandy Cove. The north shore of Puffin Island consists of rocky bluffs where an occasional puffin along with guillemots and gulls can be seen nesting among the steep cliffs.

As you leave the protection of Puffin Island, a predominant point comes into view 2 miles to the north. The shoreline remains a mixture of sheer cliffs and steep drops until you round the point and pass a gravel beach labeled on topographicals by the trig mark "Goose." The spot offers some limited camping and a small runoff stream most of the summer; it is the only possible site until you reach Garforth Island 3 miles to the north.

On the way to Garforth Island the shoreline gets even steeper and is composed of wooded terrain broken up by slabs of rock; in early June, even tongues of snow reach down to the water. If you're paddling at low tide, look for the hundreds of starfish of various colors that cling to the rock walls along this stretch. After passing several runoffs, one an impressive waterfall, you reach Garforth Island, which offers waterless camping at its north and south ends and a steep shoreline in between. The large island might also be off-limits owing to a high density of bears, but good camping conditions lie just 1.5 miles north at Muir Point.

Mount Wright

Paddling north toward Muir Point you view the mile-high mountain named after Reverend George Wright of Oberlin College. Located east of Garforth

Island, Mount Wright is known for its large population of mountain goats and as an excellent view point for much of the park. But the last few hundred feet to the summit is a technical climb that requires mountaineering equipment and expertise. Even scrambling to 2,000 or 3,000 feet of the 5,139-foot peak is a challenge, involving considerable bucking of alder and thick brush before you reach the great view points at those heights.

The traditional routes are along the creek from Dirt Glacier to the north or a steep ascent from the west across from Garforth Island. Today the route toward Dirt Glacier is a jungle of alder, so most hikers elect to begin the hike a mile or more south of Muir Point. The least amount of alder will be encountered along a steep route that begins near the stream (trig mark "Garb"), but hikers will still spend the first hour or more hacking their way through alder. Good campsites in scenic alpine meadows and ridges are found near the 3,500-foot mark northwest of the summit.

Muir Point

North of Garforth Island, the steep shoreline flattens out and the gravel beach that appears can be hiked around Muir Point toward the entrance of Adams Inlet. The point, once the site of John Muir's famous cabin, provides good camping conditions and sweeping views of the entrance to the East Arm. The area, an old gravel outwash covered by grass and dryas, is slowly being filled in by alder. Wildlife is abundant; sightings may include black bears, moose, seals, and even whales feeding 30 or 40 yards offshore. Strolling along the shoreline can be turned into a half-day hike as you can easily walk from the Dirt Glacier runoff to within a half-mile of Garforth Island, a hike of over 3 miles.

Camping sites are almost unlimited in this area except at Dirt Glacier, where willow and alder are so thick that camping on the outwash is limited to a few, well-placed small tents. The river from the glacier is swollen, even in early June, and difficult to ford without a hand line.

Adams Inlet
Round-trip from Muir Point
Distance: 21– 30 miles
Paddling time: 2– 3 days

Located 35 miles north of Bartlett Cove is Adams Inlet, an almost landlocked tidal inlet that is ringed by mountains and visited primarily by kayakers each summer. Because of the strong tidal currents and half-hidden shoals and rocks at its entrance, the heart of the inlet is an isolated spot that sees none of the cruise-ship traffic occurring on Muir Inlet. Instead, paddlers will discover a scenic hideaway and a haven for wildlife, especially waterfowl, which seems to congregate there in great numbers. Birds encountered could include arctic terns, Canada geese, loons, scoters, harlequin ducks, and mergansers, among others.

Paddlers should be aware that for several weeks around the end of July the inlet is the scene of several hundred flightless, molting Canada geese that are extremely sensitive to visitors. If kayaks get closer than a half-mile, the geese panic and do not settle down until they have moved at least a mile away. Nor will they approach campsites in the area; they tend to congregate on the outwash flats and beaches at the east end of the inlet. Paddlers should avoid disturbing the flocks of geese at all costs.

The inlet itself is a story of constant change. Nothing more than an ice field of Adams Glacier at the turn of the century, today it not only is quickly becoming revegetated but is slowly filling in, the result of uplifting land and the dumping of sediment by glacial streams. Tidal mud flats are extensive and stretch along most of the shoreline. It is best to set up and break camp at high tide, otherwise the walk down to the water is often a long and sticky one.

Hikers staying at Muir Point will find the 7-mile trek to the southwest corner of the inlet a difficult one, with much alder to bash and two major streams to contend with. Kayakers have a much easier time but still must be careful when paddling the narrows that lead into the tidal lagoon. It is a 6.5-mile paddle from Muir Point through the narrows to the island in the center. It is best to travel in on an incoming flood before high tide to reach the constricted part close to slack current. Even then, small tidal rips and whirlpools will be seen in the narrows and around a large rock marking their beginning.

Kayakers arriving late in the day can set up camp at either the Dirt Glacier outwash on the south shore or a waterless cove 1.5 miles farther east to wait for the right tidal conditions. A mile beyond the cove is a prominent rock that sits in the middle of the channel just before the constricted part of the entrance. During mid-tides the current is swifter in this section, and you will see whitewater dashing about the rock, and large whirlpools shedding from its sides. Near slack high tide, the current is greatly reduced, but paddlers still need to be careful moving through. The best route is to follow the north shore, passing between the rock and the small point labeled on topographicals by the trig mark "Upper."

Once past the rock, you'll find good camping along the south shore, where there are flats of dryas and encroaching alder and willow on what was once an old glacial outwash. The area is marked to the east by the long arm that curves into the inlet.

Many paddlers eager to enter Adams Inlet follow the 2.25-mile peninsula east toward the large central island. Here they are confronted with two channels: one heading almost due north to the north side of the inlet and the other southeast to the south side.

Adams Inlet: North Shore

The channel to the north, though often dry at low tides, quickly leads you past the huge outwash and streams of Casement Glacier. Once past the outwash you enter a narrow waterway formed by the island's north shore and the mainland. Although there is a shallow lake in the middle of the island and an occasional small runoff stream, any campsite here should be considered waterless. Alder and willow make for heavy brush across most of the island, but a small hammerhead peninsula about 1.5 miles east of the Casement Glacier outwash does have space for a campsite in its west cove.

Along the mainland you will pass several rushing streams, but there will be very limited camping because of its steep shoreline. You'll finally come to the outwash of Girdled Glacier, .5 mile wide and 8 miles from the rock in

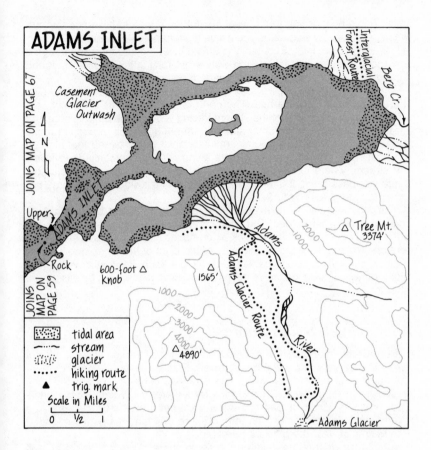

the middle of Adams Inlet narrows. At one time, some 15 years ago, there was good camping and hiking all around this outwash, but alder and thick brush have dramatically changed this. On the east side you have to search hard for a space to pitch a tent between the alder and the rocky beach. The best spots exist on the west side of the outwash.

Walking the beach south of this outwash is good even during high tide. You can follow the beach south for 1.2 miles, crossing several small runoffs, and all the time stare at the surrounding peaks, many over 5,000 feet high. If it is low tide, the extensive mud flats along the head of Adams Inlet will be criss-crossed with tracks, including those of moose, bears, and even wolves. Eventually you come to Berg Creek roaring out of the mountains. The so-called creek is a tough river to ford, especially without the help of several people and a hand line. Good campsites exist on the north side of the river, but keep in mind that the mud flats lie hidden at high tide.

Interglacial Forest Hike (Distance: 4 miles round-trip; rating: moderate): You can see a fine set of interglacial stumps by hiking 2 miles up the outwash of Girdled Glacier, and there's only a reasonable amount of alder to buck. Begin by crossing to the west side of the river and hiking up the outwash, where every step takes you a little higher and gives you a better overall view of Adams Inlet. Eventually (within a mile) the stream cuts through a low rocky moraine that is covered with alder, forcing you to hike up it and through the brush. Return to the stream at first opportunity.

The stream runs through a rock and boulder area, in the middle of which there is an alder-covered rise. The north end of this forms a narrow gap, with the steep west bank of the stream forcing you to climb up the shore and through the alder. On the other side the stream opens up to another rocky area. You'll soon spot a predominant interglacial stump over 3 feet high on the east shore. At this point you have a choice. You can continue to follow the stream, where just around the corner you'll find a fine set of stumps. Once you pass them, it is only a short way until the sides of the gorge become too steep for you to follow the river.

The alternative is to hike up the ridge along the west bank, which is marked by large landslides. The ridge may look inviting from the stream, but once you begin to scale it the thick alder makes hiking an ordeal. Those intent on climbing the ridge should stay within sight of the edge, and within a quarter-mile will be able to battle their way to a knob. It will be covered with alder, but will allow good views deep into the scenic gorge cut by this glacial stream.

Adams Inlet: South Shore

Berg Creek is only part of a huge delta formed at the mouths of a pair of rivers that curve sharply southeast toward Endicott Gap. You can begin on the south side of the delta (which has extended considerably beyond what is shown on topographicals) and hike for several miles until alder inhibits your progress. Endicott Gap is the low point between the Glacier Bay and Lynn Canal watersheds; it was through here that land mammals migrated into Muir Inlet after the glaciers retreated.

A half-mile west of the delta on the south shore of the inlet is a waterfall that has formed a small gravel fan. Camping spots exist here and would enable paddlers to spend a day away from their boats, exploring Endicott Gap. As you continue west, the shoreline becomes steep as you paddle around the base of Tree Mountain toward the gravel outwash formed by the river from Adams Glacier.

The delta is huge; by following its outer edge you first pass the channel returning to Adams Inlet narrows and then enter another passage that is formed to the north by the inlet's long peninsula. This channel takes you past a small island in the middle and to the western corner of the Adams Glacier river delta, where there are extensive slopes of dryas. The willow and alder are still sparse in this area, the campsites are plentiful, and the hiking inland is interesting. To the west is a small bay and maybe the most scenic section of an already stunning inlet.

A hike from the Adams Glacier outwash to the narrow neck of the inlet's long arm is a 2.8-mile trek one way. Most of the time you are hiking under the presence of an unnamed 4,890-foot peak that features two large snowfields on each side. Within a mile of the delta, pass a small cove and come to a glacial stream and the outwash it is beginning to form. Hike a little more than a mile up the east side of the stream to enter a deep gorge composed of ridges of soft glacial silt, almost gray and sandlike.

The outwash of the stream merges into a larger one that is like a huge sand dune. Willow and alder have already established themselves, but the hike up is easy and rewards you with more views of the gorge and the snowfields on the unnamed peak. For an excellent view of Adams Inlet and much of Muir Inlet, scramble up the 600-foot knob that lies just west of this outwash. The remainder of the shoreline begins to curve toward the neck of the peninsula and another glacial outwash now being carpeted by willow and alder. You will encounter some good campsites before reaching the final outwash. At high tide the portage across the peninsula is a .3-mile hike through an area of light brush.

Adams Glacier Hike (Distance: 5 miles one-way; rating: moderate): The easiest route to Adams Glacier is to hike along the river that carries out its meltwater and sediment, sometimes referred to as Adams River. From the west side of its large delta hike 1 mile east, following the curved lines of willow and alder that are covering the huge mounds and mesas of sediment. At this point you come to the main channel of the river and begin hiking along its west bank.

Within a half-mile, you pass a small outwash of stones with a stream leading into a group of huge sand dunes and mesa-like formations that lie

between an unnamed peak at 1,565 feet and the river. This is the start of an alternative route to the glacier, one that requires considerably more climbing. Within a mile, or maybe sooner, the river swings toward the steep bank of the west side and forces you to climb up loose sand and rubble. Any wind at all produces small, swirling dust clouds. Most likely you will be able to return to the river before having to rescale the steep ridge. This time you should be able to spot the stream to the southeast that flows from the back side of Tree Mountain.

You are forced to continue in the sand-ridge area, but the hike is enjoyable as the willow and alder are still sparse. Eventually you come to the edge and are greeted with a wide view of Adams River leading back to the glacier, which is surrounded by peaks. The glacier is still 2 miles away, but you can drop down to the river and cross the wide mud flats by first passing a stream and small outwash that lead northwest back into the sand ridges and dunes. You can follow the river another .75 mile, until you are forced onto the steep banks for the remaining mile to the glacier. Beware of "quick mud" in this section.

Another route follows the first stream and small outwash you come to along the west bank of Adams River. Hike it to the southwest as it climbs into the middle of some sharp-edged hills of sediment. When you approach a fork with another small outwash, stay to the left; in 1.6 miles from Adams River the stream and rubble curve to the west. Walk 100 yards or so due south and you'll come to the edge of a ridge with an incredible view from the top.

Below is a stream and canyon that lead back to Adams River. You can descend into the canyon and follow the stream to the river. Or after reaching the stream, you can cross it and climb the south ridge of the canyon. From the top of the south side, make your way to the mountain slopes that form the west side of the Adams River valley. Hike the mountainside to stay well above the river and mud flats. You can reach 1,200 feet or even higher on the barren slopes before dropping back down to Adams River and near the face of the glacier in 2 miles.

Muir Point to Nunatak Cove

Distance: 12 miles
Paddling time: 5–7 hours

For those paddling straight to the glaciers of Muir Inlet, caution has to be exercised when crossing the mouth of Adams Inlet. From Dirt Glacier outwash to Point George, it is a 1.5-mile open-water crossing that can get choppy at mid-tides or during a period of wind from any direction. After rounding Point George you come to Maquinna Cove and a smaller cove just north of it with limited places to land and few campsites. From here the next 1.6 miles of shoreline are steep as you round Klotz Hills, which really appear as one.

Eventually you round a small point (marked on topographicals by the trig mark "Cush") and then come into view of a huge outwash. Looking beyond it to the northeast you'll see Casement Glacier. For those paddling up bay from Bartlett Cove this is often a special moment, as Casement is usually the first glacier they spot after three or more days on the water.

This outwash is huge (it's a mile-long paddle to round it) and is quickly being covered by willow and alder. Those who stop will discover some large tidal pools behind a rise of gravel and stones on the beach, and many good campsites at its north corner, where a strip of dryas separates the advancing alder and the rocky outwash. The large outwash also marks the beginning of a 7-mile beach and some excellent beach hiking that begins north of Klotz Hills and ends at the back of Goose Cove.

Paddling north from the large outwash you will spot the sharp ridge that borders the north side of Forest Creek along the east side of Muir Inlet, and behind that the rounded peninsula that separates Goose and Nunatak

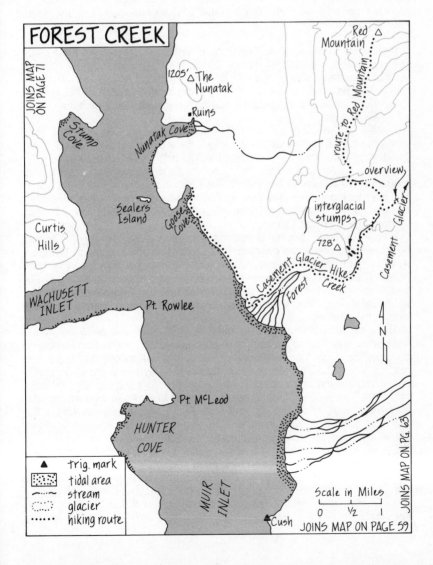

FOREST CREEK

JOINS MAP ON PAGE 71

Red Mountain

1205' △ The Nunatak

route to Red Mountain

■ Ruins

Stump Cove

Nunatak Cove

overview

interglacial stumps

Sealers Island

Goose Cove

Casement Glacier

Curtis Hills

728' △

Casement Glacier Hike Creek

Forest

WACHUSETT INLET

Pt. Rowlee

N

Pt. McLeod

HUNTER COVE

▲   trig. mark

tidal area

stream

glacier

hiking route

MUIR INLET

Scale in Miles

0   ½   1

JOINS MAP ON PG. 63

▲ Cush   JOINS MAP ON PAGE 59

coves. To the west is Curtis Hills, which form the north shore of the entrance to Wachusett Inlet. Keep in mind that distances look deceivingly short. It is still a good 5-mile paddle from the outwash past Forest Creek to Goose Cove, though from the water it looks like less than 2 miles.

It is near Goose Cove that many paddlers spot their first iceberg, and these are only the largest ones; they sail toward the ocean but they rarely make it. The cove, once the site of a backcountry ranger station, is a scenic spot, with The Nunatak rising right behind it and Red Mountain to the northeast. Goose Cove is usually free of icebergs during the summer and offers excellent campsites along the east shore. Water (that the park service recommends be treated for giardia, an intestinal parasite) is available at the head of the cove, from where a natural overland route once existed to Nunatak Cove to the north. But a jungle of alder has made that mile-long hike a brutal trek.

Casement Glacier Hike (Distance: 8 miles round-trip from Muir Inlet; rating: moderate to difficult): Forest Creek, the traditional route for a close view of Casement Glacier or a hike up Red Mountain, remains the best one today. But alder and willow are quickly filling in the outwash of the stream (once the drainage of the glacier), making it necessary to buck a little more alder every year.

If starting from Goose Cove, add another 2 miles each way. The trek along the beach is pleasant, especially if the weather is clear, as you'll be greeted with stunning views of Mount Wright and Mount Case to the south. When you approach the ridge on the north side of the outwash, keep hiking a quarter-mile before turning inland. The brush is thick along the ridge; you'll find that the most open areas are along the main channel toward the center of the outwash. Thick alder will be encountered within a half-mile; it is best to stay along the north shore.

Where the brush begins to thin out, you pass a pair of canyons to the north, with the one to the east formed by a hill with an elevation of 728 feet. At this point there are several natural routes within view, but remember that Forest Creek swings southeast here, passing all the steep-sided canyons and running along the south side of hill "728." As you approach this hill, the alder and willow become thick again, so it is best to cross over to the south side of the stream.

The alder remains dense for a half-mile or so, then you break out into a wide-open area that is marked by at least one green pond. The walking is considerably easier here as Forest Creek swings to the north and continues circling hill "728." The east side of the hill is partly loose gravel and rock containing a stand of interglacial trees. A couple more trees lie in the middle of the stream. Follow Forest Creek for another half-mile to where there are sections of soft, boot-sucking mud and, eventually, a waterfall with a 12-foot drop along solid rock. Alder and brush get thicker but are interrupted in places by some interesting pools that lie in rock basins.

Eventually a small stream flows into Forest Creek, though at times it may be dry. Separating the two streams is an east-west ridge with a high point at its east end and sheer, loose rock on its south side. Brush is thick along it but you can scramble up for a view of the fast-retreating Casement Glacier to the north, Adams Inlet to the south, and the lower end of Muir Inlet to the southeast. To reach the glacier, follow the ridge until you drop down into

Casement's large outwash, which extends south to Adams Inlet. Head north, climbing loose moraine of rock and shale for an additional 2 or 3 miles. Casement is a stagnant glacier, and the area around its terminus is characterized by black ice, many ponds, and soft mud.

Red Mountain Hike (Distance: 8 miles one way from Muir Inlet; rating: difficult): One route to the summit of this 3,580-foot mountain is to continue along Forest Creek as it approaches hill "1268" of the Casement Glacier Hike (see above). Follow the east side of hill "1268," heading north through extremely thick alder and willow; you have to labor to get through the jungle of brush for the next 1.5 miles. Eventually you approach a stream that leads west to a low point between hill "1268" and the southern ramp of Red Mountain. Battle your way up to the low point and continue north on to Red Mountain, where the brush thins out as you ascend. From here the alpine walk is moderately easy to the summit, with few steep grades—but remember that snow persists on the mountain well into August. The hike to Red Mountain makes for a long day; some hikers shorten it by camping near the green pond of Forest Creek rather than Goose Cove.

Nunatak Cove

Nunatak Cove and The Nunatak, a 1,205-foot knob, are reached after a 1.5-mile paddle north from Goose Cove. As you pass Sealers Island, the shoreline is rocky and sheer but quickly gives way to a shingle beach that extends into the cove. This is another scenic spot, as there is usually a sprinkling of icebergs along the cove's shoreline, which is dwarfed by The Nunatak to the north. The best camping is on the south shore, which is composed of shingle and rock. Almost the entire back half of the cove is an extensive mud flat with a shoreline of soft mud.

Emptying into the head of the cove is a stream unnamed on USGS maps but often referred to as Nunatak Creek. On the other side of the stream, in the northeast corner of the cove, are the remains of former mining activities. Overshadowing everything is The Nunatak. From its summit you get an extensive view of Muir Inlet, Red Mountain to the east, and Adams Inlet to the south. Ten years ago this was a common hike, but alder, the hiker's nightmare, has taken root and turned a scramble up the knob into a tough afternoon of battling branches. Alder now covers the conventional route from the south and most of the top except the rocky point at "1205." Those who want to tackle this route should plan on spending 3 to 6 hours round-trip and begin by crossing the creek, no easy task in July or August. The best place to cross is a half-mile inland; from here the route up is quite visible, as you can view a small brush-covered knob to the south, the main rise of The Nunatak to the north, and a ravine between them. Hike toward the ravine and gradually swing toward the main peak to the north. There are dozens of routes up and they all involve grasping alder branches and skirting around rock ledges and faces.

As you near the summit, the steep sides of The Nunatak give way to dips and flat spots filled with brush. The first view is of Muir Inlet to the south, but from the rock summit you are at last rewarded with a 360-degree panorama of the East Arm and its glaciers spilling out of the surrounding mountains. Another route is to climb the north side of The Nunatak from outside the cove. Although there is less brush, the route up is extremely steep along ledges and sides of rotten rock.

McBride and Riggs Glaciers
Nunatak Cove to Riggs Glacier
Distance: 8 miles
Paddling time: 3–6 hours

If you follow the east side of Muir Inlet, it takes almost 1.5 miles to paddle past The Nunatak. But once beyond it, you will view a shoreline of eroding ridges, gorges, and mesa-shaped formations. To the west is Wolf Cove (see chapter 9) and beyond it the sheer east side of White Thunder Ridge. On the east side two streams, and the deep gorges they have cut, lie up bay from The Nunatak and offer good camping spots and avenues inland— especially the second outwash to the north, where a winding ravine can be followed to meadows of dryas, light alder and willow.

But a hidden glacier lies ahead. Many paddlers pass by this area and head toward Van Horn Ridge and its prominent peak of 2,124 feet in anticipation of their first tidewater glacier. After rounding two points, the second labeled on topographicals by the trig mark "Carl," you are greeted with the stunning sight of McBride Glacier and a collection of icebergs along the shoreline. At high tide the ice is crowded within the lagoon that lies in front of the glacier, but by low tide the bergs spread out into Muir Inlet.

Camping is possible along the beach 100 yards or so north of the lagoon. There you can pitch a tent on dryas among the encroaching alder. Water can be gathered from numerous streams running off McConnell Ridge, and the bergs offer an unlimited supply of crushed ice at your doorstep. You can

Riggs Glacier in Muir Inlet

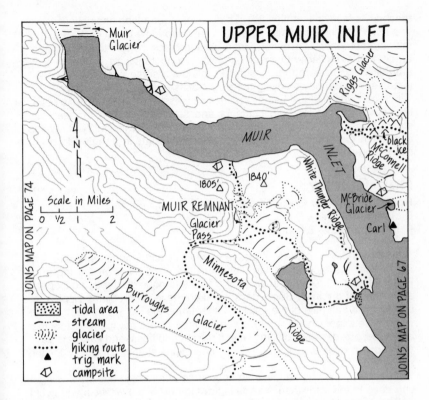

hike a half-mile or so along the base of McConnell Ridge toward the glacier, but beware of falling rocks, as the ridge is steep and loose. No one should think of pitching a tent in this section.

It is possible to hike along the beach at the base of McConnell Ridge to a stream with a small gravel outwash a mile north of the glacier. In early summer the stream lends itself well to a 0.5-mile scramble up McConnell Ridge for a good overview of the area but, unfortunately, no views of Riggs Glacier. The shoreline at this point turns steep and prevents most hikers from continuing their casual walk.

In a kayak, you continue north and round several small points that form the western end of McConnell Ridge. Along the way, Riggs Glacier gradually reveals itself. It has now receded into a small lagoon of its own, with a gravel arm separating the water in front of the face from Muir Inlet. The arm, which allows you to pitch a tent with the face of the glacier looming above, is viewed as a popular spot by backcountry users and an overused one by the park service. At one time the local tour boat *Thunder Bay* dropped off most of its campers here, but stopped doing so in 1987. When staying here, make sure all tents are pitched above the rising tide and the reach of the large waves that could be generated by a sudden calving.

Riggs, once an active glacier, is largely grounded now, as the sea reaches its face only at high tide. The glacier is retreating and now consists of a main face and a smaller arm that wraps around a rocky knob. South of this and bordering McConnell Ridge are the black-ice remains of a former arm of McBride Glacier that is still shown on topographicals. There are several hikes in the area (see chapter 9), although some people are content just to spend the day studying the blue cracks of Riggs' towering face.

In recent years Riggs' beauty and easy accessibility have made it a popular gathering spot for both campers and kayakers, and on a clear day steady lines of small sightseeing tour planes loop past it. The park service eliminated the area as a drop-off point for tour boats in an effort to begin site restoration.

Muir Glacier
Round-trip to Muir Glacier from Riggs
Distance: 16 miles
Paddling time: 5–8 hours

It is an 8-mile paddle to the face of Muir Glacier; since campsites are rough near the face, most kayakers turn the paddle into a day trip from Riggs Glacier. The key to seeing John Muir's famous glacier is the icepack at the end of the inlet. Some days kayakers have no problems zipping up bay within a half-mile of the glacier. Other times the ice is so thick that it is impossible to get close enough to see the face. There are many factors that affect icepack density including wind direction and tides, and kayakers have little control over any of them since most need almost a full day for the round trip.

If possible, it is best to leave an hour or so before high tide. This will put you deep in the arm when the tide is flowing out, which will help you avoid getting trapped in the icepack. The incoming tide tends to condense the ice and often will shut off channels and passages that were open three or four hours earlier. The best routes are along the shoreline, where meltwater streams emptying into the inlet push the ice away.

By following what starts off as the south shore, you will see the glacier sooner from a greater distance away. This entails departing from Riggs and paddling toward the northern point of White Thunder Ridge, almost a 1.5-mile crossing. The shoreline will be steep and rocky as you pass the north end of White Thunder Ridge, but within 2 miles there is a large meltwater stream forming a small outwash. There are some rough campsites suitable for a small tent above the stream on either side. A mile up the stream is a low point between two knobs, one 1,805 feet in elevation to the west and the other 1,840 feet to the east. Scramble up either knob for good views of upper Muir Inlet. The pass and the stream leading to tidewater make up the final leg of a 6.7-mile route from Wolf Cove (see chapter 9).

The shoreline continues west for another 1.5 miles before it begins to curve to the north. At this point the inlet will most likely resemble a sea of ice populated by sun-bathing seals. Observant kayakers can hear the occasional cry of a lost pup above the constant crackling of the ice. The shoreline is steep and barren along the west side of the inlet but is broken up by an occasional stream gully that allows you to climb to an overview of the area. Few, if any, of these ravines have enough level space to accommodate a tent.

Waterfowl in Geikie Inlet

Four miles from the glacier the inlet begins to curve north, and from there a portion of the glacier can be seen in the mountains. Kayakers will have to paddle another 2 miles before they can spot the face of this active tidewater glacier. The west shore is extremely steep but does provide the only remote place to camp, with a view of Muir Glacier from your tent door. A mile from the face of the ice is a stream and small gravel fan. South of the stream there is space for one, maybe two, small tents, giving you a postcard view of Muir Glacier at the cost of numerous rocks under your sleeping pad.

The east shore at the head of Muir Inlet is marked by a point with two distinct knobs 1.5 miles from the glacier. Kayakers following this shoreline will have to round the point before they can view the face. Suitable campsites exist south of the point along two large outwashes and streams. You can't see Muir Glacier from the shoreline, but you can scramble up a small stream to the first knob on the point for a view of it.

Wachusett Inlet
Round-trip to Carroll Glacier
Distance: 26 miles
Paddling time: 2–3 days

Fifty years ago Wachusett Inlet was little more than a small dent in the ice of Plateau Glacier. Today it is a 13-mile inlet, free of ice, and Carroll Glacier, not Plateau, forms the head of the waterway. Such is the constant change of Glacier Bay.

Although the glaciers have receded, and alder and willow are galloping up

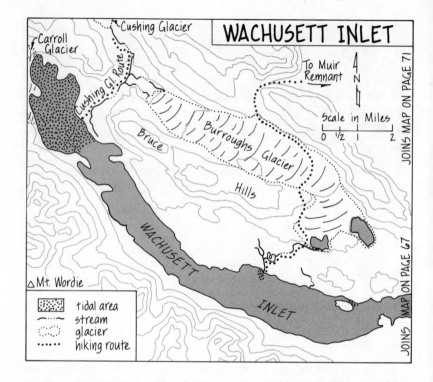

the inlet, Wachusett still offers open terrain to hike, ice remnants to explore, and a bit of solitude for the kayakers who spend two or three days paddling and camping this arc of a waterway. The mouth of the inlet is almost due west of Forest Creek and is a 9-mile paddle south from Riggs Glacier or an 8-mile paddle north from Muir Point. From here Wachusett curves more than 13 miles to the northwest to end at Carroll Glacier, which now has receded onto land.

The north shore is mostly glaciated moraines left by the once-mighty Plateau and Burroughs glaciers, and the south shore is a much steeper and rougher coastline with intermittent gravel beaches and outwashes. There are good campsites at the mouth on the south shore, but neither side offers too many more the first 2.5 miles in, owing to steep shorelines and thick brush. Along the north shore, you pass a peninsula in 2.5 miles, one that turns into an island at high tide. One mile west of the peninsula, dryas-covered outwash with a small stream flowing through it is a good campsite. Across the inlet is impressive Idaho Ridge with its hanging glacier and 3,500- and 4,000-foot peaks. This scenic spot is also the beginning of one route to Burroughs Glacier Remnant (see chapter 9).

There are several more stream outwashes and fans on the north shore suitable for good camping in the next 2 miles before the inlet begins to curve to the northwest. Once around the corner (a 5-mile paddle from the peninsula), you can view the end of the long inlet and see the white surface of Carroll Glacier, though Wachusett Inlet often generates a fog and wind of its own that might obscure it. At this point Bruce Hills begin to angle toward the water and dominate the north shore.

Among the better camping spots in the upper half of the inlet is one on what is now the western shore, 3.5 miles from the end or 1.5 miles from the first of three distinctive spits (not shown on the topographicals). Here several peaks come together to form a bowl that leads gently down to the Inlet. Several streams cut into the gentle grade and form a small fan along the shoreline, where there is enough space for a couple of small tents. From this spot campers can trek a half-mile into the bowl to view a thundering waterfall.

The first spit lies on the now western shore, the second, a half-mile beyond on the eastern shore. It is near these small peninsulas that the best camping is found at the head of the inlet. From the first spit to beyond the head of Wachusett Inlet, the shore is a vast mud flat at low tide and is best approached only during high tide. Otherwise it is a long trudge through the soft mud to solid ground. Carroll Glacier now has black-ice mounds at its terminus, having retreated onto land along the western shore. The opposite side of the inlet is a huge gravel delta formed by the rivers from Cushing and Burroughs glaciers, with mud flats surrounding it.

Cushing Glacier Hike (Distance: 3 miles one way; rating: moderate to difficult): The hardest part of this route is getting to the north side of the river that spills out from Bruce Hills. It is best done at high tide in a kayak, as the river runs deep and even in early summer is hard to ford. Hiking along its south side is considerably harder as you contend with a much steeper slope and more alder and willow.

From the north side follow the gravel fan back until the river swings into a cliff, forcing you to make a steep ascent onto the ridge. The terrain on top is open, with many small gravel moraines that are covered by dryas and scattered alder and willow. After a half-mile, you trek over a hill and then begin a steady climb, with the river and gorge in sight below. Eventually you pass where the rivers from the glaciers join together to form the main channel.

At this point you begin to swing north, keeping in sight of the west shore of the river leading to Cushing Glacier. Within a mile climb a small ridge and then begin to climb a second mound at 700 feet, about 3 miles from the inlet. From here the views of the glaciated valley are good, although Cushing Glacier still lies almost another mile to the north.

Lower Muir Inlet
Wachusett Inlet to Tlingit Point
Distance: 16 miles
Paddling time: 1–2 days

Point Rowlee is the south side of the entrance to Wachusett Inlet and the beginning of lower Muir Inlet's eastern shore, an area of many beaches and few paddlers. The 16-mile paddle to Tlingit Point can be accomplished in one day with the help of the tides or easily in two, with time to spare for exploring the long beaches along this stretch. The first beach lies between

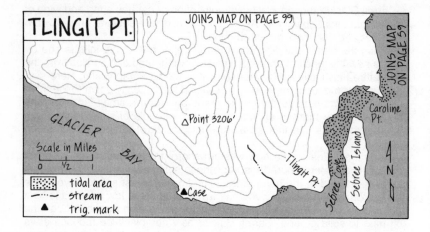

Point Rowlee and Point McLeod, just before paddlers begin turning into Hunter Cove, the site of the second. The cove has a backdrop of Idaho Ridge to the west and provides camping possibilities in its northwest corner. If you don't mind bucking some heavy alder, the southernmost river emptying into the cove provides the best route up Idaho Ridge.

It is a 1.5-mile paddle across Hunter Cove and then you come to a shoreline of several small points and gravel beaches in the 4-mile stretch to the outwash of Morse Glacier. The river from the glacier and the shingle beaches on either side of its outwash provide good campsites and lots of driftwood for campfires. It is a tough trek along the creek, which is overrun by alder and thick brush, for anybody wanting to reach either the stand of interglacial wood located along it or farther beyond, the glacier itself. Remember that interglacial wood is protected within the park and should never be removed or used for campfires.

More shoreline campsites and good beach hiking can be found at Caroline Point, another 4 miles south of Morse Glacier outwash. You'll spot the distinctive point several miles before reaching it and even view the ridge of Sebree Island behind it. Caroline has long stretches of gravel beach on both sides of its point, which usually contain much driftwood, and is frequently visited by black bears.

Caroline Point also marks the beginning of the channel between the mainland and Sebree Island. To paddle it you have to approach it less than an hour before or after high tide. By paddling through, you quickly round Tlingit Point and the cove just west of it. This is a spectacular spot to camp, since you will have the Fairweather Range to the west, Mount Wright to the east, and the entrance of Glacier Bay, what seems like miles of water, in front of you. The stream emptying into the cove is clear and cold. There are lots of strawberries in July and an abundance of driftwood as well. Camp at this crossroad of Glacier Bay on a clear night and you will undoubtedly feel the immensity of the park.

6 WEST ARM

The West Arm is viewed by most as the larger, more spectacular of Glacier Bay's two main inlets and traditionally the isolated corner of the park. For years its beauty was accessible to kayakers only after costly float-plane drop-offs or a long haul from Bartlett Cove. Many paddlers, with either a time or a budget crunch, elected to experience Muir Inlet and leave the wonders of Johns Hopkins and Tarr inlets to their imagination.

But in 1985 the park's tour boat began a drop-off service from Bartlett Cove, and kayakers were suddenly provided easy transport to upper portions of the West Arm. The initial places were Reid Inlet and Lamplugh Point, just east of the glacier of the same name. Both spots are only a day's paddle from Johns Hopkins Inlet, and the immediate result was a concentration of kayakers around Reid Inlet and the Ptarmigan Creek area while the rest of the arm continued to bask in solitude. But this might change in the near future, as the NPS plans to rotate the sites that draw large numbers of visitors.

The stunning tidewater glaciers of Reid, Johns Hopkins, and Tarr inlets in the northwest corner of the arm are the destinations of most paddlers. But the other fiords and inlets offer their own unique and rewarding scenery, glaciers, and hiking opportunities. On the east side of the West Arm lie Tidal, Queen, and Rendu inlets, while at the southwest entrance are the many corridors of Hugh Miller Inlet.

Those paddling up bay from Bartlett Cove should plan on spending three to four days to reach Hugh Miller Inlet, a one-way trip of around 38 miles, which involves crossing the bay north of Willoughby Island. Reid Inlet lies another 17 miles to the north or a good five- to six-day paddle from the park headquarters. The average kayaker, who covers 10 to 12 miles a day, needs at least 14 to 18 days to paddle round-trip and explore the West Arm, including the inlets along the east shore.

CROSSING THE BAY

Kayaking across to the west side of Glacier Bay is an open-water paddle that should be done with extreme caution and during calm conditions. The shortest route would be to paddle from Strawberry Island to Berg Bay, but this involves the Sitakaday Narrows, a channel that produces heavy swirls and unpredictable rips during tides and at times reaches a velocity of 6 knots. The same holds true for the narrow section between Young Island and Rush Point on the west side and for any section south of that to Icy Strait.

Paddlers should not attempt to cross Sitakaday Narrows, as even vessels more than 100 feet long have been spun around by the unpredictable currents. The park service recommends that kayakers not cross the bay anywhere south of Flapjack Island. This small barren island marks the north end of the Beardslee Islands and is a day's paddle out of Bartlett Cove. The best bet is to camp near Flapjack and then in the calm of early morning—if conditions and tides are right—paddle the 4.5 miles to Johnson Cove on

Willoughby Island's northeast corner. You should plan to leave the area of Flapjack an hour before high tide, as it is the ebb tide that produces the strongest currents. Johnson Cove makes for a scenic campsite with a protected harbor but little if any water available. From the cove, it is a 3-mile trip to the mainland north of Fingers Bay.

A longer but safer alternative is to paddle to Muir Point, a two- to three-day trip out of Bartlett Cove, and then make the 1.5-mile crossing to the shore north of Caroline Point. From the point it is a 12-mile paddle to Tidal Inlet on the east side of the arm and another 2 miles of open water to Hugh Miller Inlet.

Hugh Miller Inlet
Geikie Inlet to Scidmore Bay Tidal River
(Not including Charpentier Inlet)
Distance: 18 miles
Paddling time: 2 days

Hugh Miller Inlet is a scenic side trip from Glacier Bay's Lower West Arm and a hideaway where you can spend days exploring its many fiords, bays, and coves, all set to a backdrop of mountains and snow-covered peaks. Paddlers will usually find calm water even when the bay itself is rough, an

endless number of good campsites, and numerous glacier valleys to explore. The glaciers in this area have long since retreated onto land, and although they do not have the dramatic features of their tidewater cousins, they are still impressive.

The area can be entered from the north through a tidal channel between the West Arm and the northern end of Scidmore Bay, though only at the highest tides does the channel completely connect. Or you could enter the inlet from the east through a channel among the islands of Blue Mouse Cove. To the south, Charpentier Inlet reaches within a mile of Geikie Inlet, but the overland portage, at one time a shortcut that would have saved miles of paddling and backtracking, is no longer a feasible hike, especially if you have to buck the alder with a boat over your head.

From the north side of the entrance to Geikie Inlet, it is 7 miles to the entrance of Hugh Miller Inlet in what, for the most part, is a paddle around the 2,765-foot Hugh Miller Mountain. The shoreline is steep most of the way, but you do pass two streams, the first being 2 miles north of Geikie. This stream is impressive as it cuts through a rocky bluff in a series of waterfalls and leaps before emptying into the bay at a narrow gravel strip. The second stream is around the next point; it also makes for a good place to land.

The entrance to Hugh Miller will be sighted in another 3 miles when you

Hugh Miller Inlet from Inner Sanctum Knob (Ed Fogels photo)

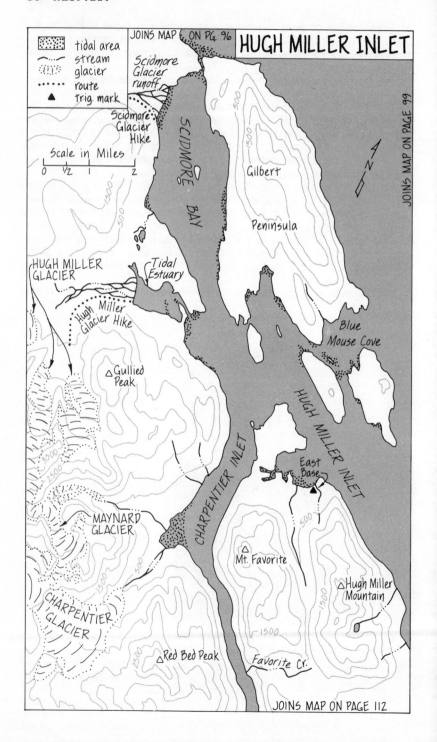

HUGH MILLER INLET

tidal area
stream
glacier
route
trig. mark

JOINS MAP ON PG. 96

Scidmore Glacier runoff

Scidmore Glacier Hike

Scale in Miles
0   ½   1   2

SCIDMORE BAY

Gilbert

Peninsula

HUGH MILLER GLACIER

Tidal Estuary

Hugh Miller Glacier Hike

△ Gullied Peak

Blue Mouse Cove

MAYNARD GLACIER

CHARPENTIER INLET

HUGH MILLER INLET

East Base

△ Mt. Favorite

CHARPENTIER GLACIER

△ Red Bed Peak

△ Hugh Miller Mountain

Favorite Cr.

JOINS MAP ON PAGE 99

pass an unnamed cove to the west. To the north are the islands that make up Blue Mouse Cove, with the south end of Gilbert Peninsula looming overhead. Both coves provide good camping and calm water, but Blue Mouse is often the favorite. The cove, supposedly named after a well-known theatre in New York, features shingle beaches, protected water, and pleasant views across the bay to the east; and on a clear day you can see Mount LaPerouse, 10,728 feet in elevation, to the west. An especially nice campsite in Blue Mouse is the small cove at the northwest end of the large island to the east.

The best camping area in the unnamed cove is near the stream (trig mark "East"), where there are fewer tidal flats to deal with. Throughout this cove, as in much of Hugh Miller Inlet, you will spot moose droppings and prints and perhaps even the animal itself. The sandy banks of the central stream in this cove are a good place to look for moose, bear, and wolf prints.

Charpentier Inlet

It is a 9.5-mile paddle from Blue Mouse Cove to the head of Charpentier Inlet, a narrow, steep-sided fiord that makes for an interesting side trip. The first 4 miles from its entrance heads almost due south along a waterway that is .75 mile wide and features shingle and gravel beaches on its east shore. This section ends at a large outwash and tidal-flats area on the west side, where the meltwater streams of Maynard and Charpentier glaciers empty into the inlet. Many paddlers camp along the edges of the gravel fan, then explore the rest of Charpentier in an empty boat. The two glacier valleys also provide opportunities for day hikes, though alder has now completely covered the beginning of the route to Maynard. By skirting the ridge south of the Maynard outwash, you can trek back into the valley and avoid much of the alder.

Beyond the outwash, the inlet curves to the southeast and narrows to a .25-mile passageway while becoming considerably steeper and more impressive. In the beginning of the 4-mile corridor, you paddle between the flanks of Mount Favorite to the east and Red Bed Peak to the west along a shoreline that is now being reclaimed by brush and alder. There are few places to land in the first 3 miles, and the only camping spots are a few limited ones near Favorite Creek, 1.3 miles from the head, and at the end of the inlet, where you enjoy an immense view up Charpentier Inlet.

Hugh Miller's Inner Bay

North of Charpentier and just south of the narrow entrance of Scidmore Bay, there is what appears from a kayak to be a mile-wide cove on the west shore, with an extensive pebble shoreline in the center. Though at times it is difficult to see the opening from the middle of Hugh Miller Inlet, there is a 0.8-mile-long tidal waterway that leads into an inner tidal estuary. The spot is unique, with the flanks of Gullied Peak bordering it along the south and the extensive outwash of Hugh Miller Glacier on the west, and it's a haven for nesting geese, ducks, and other waterfowl. Often you can enter this area and see what looks like thousands of birds taking off, landing, and bobbing on the calm surface. Most of the waterfowl will be molting, kayakers should stay at least a half-mile away to avoid harassing them.

Every six hours the tide flushes in or out of this bay; catching the right tide is the key to visiting the area. Enter or exit the bay either with the direction of the tide or wait until slack tide. The current rushing through the narrow opening is strong, often too strong for most kayakers to paddle against, and creates small whirlpools and rips along the way. It's best to time your arrival and departure with the tide, as it is extremely difficult to line your boat along the steep banks.

In the estuary the water is calm and the bird life plentiful. Good camping spots abound on both the north and south shores toward the west end, where the alder hasn't yet reached jungle stage. The north side is composed of wide dryas flats, but note that water may be a half-mile walk away at times. The south shore features gravel benches and numerous runoff streams from the Gullied Peak ridge. At the very back are the extensive outwash and tidal mud flats of the Hugh Miller Glacier, which is slowly filling in the estuary, day by day.

Hugh Miller Glacier Hike (Distance: 3 miles one way; rating: easy): The trek to the glaciers at the end of the outwash can be attempted on either side of the main channel of the river, but the south side has fewer meltwater streams to cross. From the southern corner of the tidal flats, hike up the outwash, following the edge of the ridge. Often it is necessary to climb up the ridge and alder to work your way around the steep bank of one of the channels. There will be numerous small streams and channels to cross in the first 2 miles, as you head for the knob where the glacial valley curves to the south.

Within 2 miles you begin to work around the ridge. The vegetation thins out considerably and an occasional large log of interglacial wood appears. You might also notice the number "4" painted high on a rock along the way; this is an old photo station where researchers used to record the retreat of Hugh Miller Glacier. Once around the knob, you spot the terminus of one glacier sloping down, and high above it to the west are the remains of Hugh Miller Glacier. In this stretch, it is easy to envision the flowing ice pushing its way north and forming the moraines that remain on each side of the valley. Scrambling up the ridge of Gullied Peak to the east is easier at this point and provides an excellent overview of all the glacial ice that is left at the head of the outwash.

After turning the corner of the knob, you can continue another half-mile south and reach a major stream thundering down from the east. To ford it in early summer, requires a hand line. By mid-July the stream's torrential speed and depth may well make a crossing impossible for most hiking parties. Likewise hiking on the other side of the glacial valley, major branches of the river stop your progress when you are still more than a half-mile from the glacial ice.

Scidmore Bay

Scidmore is the northern arm of Hugh Miller Inlet and a well-paddled waterway offering calm passage to kayakers traveling to or from Reid Inlet. The entrance is a narrows .5 mile wide. It soon opens up to an impressive body of water measuring 5 miles in length and is 1.6 miles at its widest point, near a pair of distinctive islands in the middle. Follow the west shore

Hiking toward Scidmore Glacier (Ed Fogels photo)

of the narrows to pass some high rock bluffs with glacier tracks streaked across them horizontally.

The shoreline of the islands is steep, making landing hard. The same is true for much of the east shore, especially the northern portion, as its steep, rocky slope offers few places to camp. The west shore is composed of a long gravel beach, once you pass the rock bluffs and reach a cove and stream 2.3 miles from the entrance. The stream flows from some small lakes to the south, but the 0.6-mile trek to them is a fierce battle with alder.

The best place to camp, and one of the most scenic in Hugh Miller Inlet, is at the north end of Scidmore on either side of the outwash of Scidmore Glacier. On the south side there are stretches of dryas-covered gravel that support only intermittent alder. After a short hike up the outwash, you are greeted with a view of the glacier spilling out of the mountains to the west, all of Scidmore Bay to the south, and Queen Inlet across the main bay to the northeast. Camping spots are plentiful north of the outwash, right to the tidal channel.

Scidmore Glacier Hike (Distance: 2.5 miles one way; rating: moderate): Although the glacier is visible from almost any spot in the north end of the bay, a trek up its valley takes you to the fringes of the granite bowl into which it spills. Either side of the main channel can be hiked, but the dryas plain on the south side makes for a more pleasant stroll in the beginning.

Eventually the dryas gives way to the barren rocks and gravel of the outwash, which can be followed for a half-mile before high banks of loose gravel and morainal debris begin to close in. Where the river takes a swing to the southwest, scramble up a low shelf of gravel and boulders, cross a small stream, and continue toward the bowl, with an excellent view of the glacier straight ahead. In another half-mile or so, you'll be faced with a major stream flowing north into the main channel. In early summer, there is no problem in crossing it, but in late July or August you might need to search for a safe place to ford.

Not long after you've forded the stream, the high banks of gravel close right in on the main channel, forcing you to climb up the loose mounds. Once on top, you'll find a hilly and dryas-covered route that can be followed for a half-mile (or even farther west) right into the bowl of Scidmore Glacier, until the snow or steepness of the terrain turns you back. Up close, Scidmore is a beautiful glacier whose snout drops sharply out of the mountains. Its ice is clean but rugged, showing off a spectrum of blues and whites.

Reid Inlet
Scidmore Bay to Reid Inlet
Distance: 8 miles
Paddling time: 3–5 hours

From the north end of Scidmore, there is a tidal channel. Here you can paddle into the main bay for a shortcut out of Hugh Miller Inlet. At high tide, you may be able to travel it without ever getting out of your boat, but most kayakers are forced to climb out and haul their crafts 50 yards across a gravel ridge at the south end. Hit it at low tide, and you either walk your boat halfway down the channel through soft mud or take an unscheduled 2-hour break.

Emerging in the West Arm kayakers heading north will soon see their first icebergs. On a clear day there is a good view from this shore to the northeast, as you can see the entrances of Queen and Rendu inlets along with Composite Island and a scattering of bergs, all shapes and sizes, floating down bay. The best camping spots on this rocky shoreline are west of the tidal channel.

The shoreline to Reid Inlet is steep most of the way, but you do pass two gravel beaches. The first is 2 miles to the north, where a major stream empties into the bay. The second is another 3 miles beyond and appears as a .5-mile beach with a stream and small river flowing into it. This strip is a scenic setting for a camp and well worth a night spent. It is marked to the east by a distinctive knob and to the west by a ridge that is snowcapped until mid-summer. The view is extensive from the shoreline, as you can see Russell Island, Mount Abdallah, and the Grand Pacific Glacier to the north, and the Fairweather Range to the northwest. The best camping is along the western edge, as there is a large tidal flat fronting most of the shore.

It is also possible to trek up the old glacial valley that lies between the knob and the ridge. Although the alder and willow are getting thick, you hike along the base of the ridge with a minimal amount of brush to contend with, and in 1.5 to 2 miles you get to view the river flowing through an open outwash that stretches to the southeast. This outwash makes for an interesting day hike. In 2 miles it begins to curve into the mountains to the southwest.

Once past the beach, you follow the steep shoreline around Ibach Point to enter Reid Inlet, whose highlight is the tidewater glacier of the same

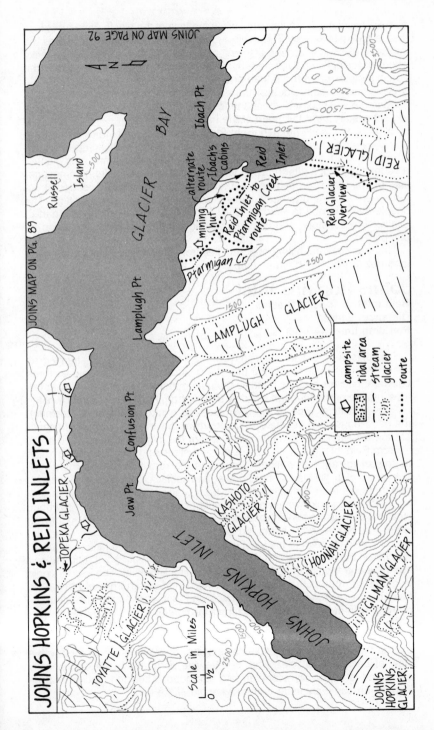

JOHNS HOPKINS & REID INLETS

JOINS MAP ON PG. 89

JOINS MAP ON PAGE 92

Russell Island

GLACIER BAY

Ibach Pt.

Ibach's cabins

Reid Inlet

alternate route

mining hut

Reid Inlet to Ptarmigan Creek route

REID GLACIER

Reid Glacier Overview

Ptarmigan Cr.

Lamplugh Pt.

LAMPLUGH GLACIER

Confusion Pt.

Jaw Pt.

TOPEKA GLACIER

KASHOTO GLACIER

HOONAH GLACIER

GILMAN GLACIER

JOHNS HOPKINS INLET

TOYATTE GLACIER

JOHNS HOPKINS GLACIER

campsite
tidal area
stream
glacier
route

Scale in Miles

0   ½   1   2

name. The inlet is 2.5 miles long, while the face of the glacier measures less than a half-mile across. The river of ice is beautiful but not as active as the Johns Hopkins or Margerie glaciers in Tarr Inlet. Still it rumbles and cracks often, occasionally calves and, by filling the inlet with ice, has delayed the floatplane pick-up of more than one party. There is good camping near the gravel spits on both sides of the entrance, though most paddlers tend to choose the west corner where Joe Ibach's cabins still stand. Limited camping spaces are near the head of the bay.

In 1925, Joe Ibach put ashore at Ptarmigan Creek and struck a gold vein nearby. For the next 30 years, he and his wife, Muz, traveled up bay each summer to work the claims. In 1940, they built their cabins, and Muz planted a vegetable garden after hauling dirt up in empty ore sacks from Lemesurier Island. The three spruce trees that form the backdrop to the wooden structures were also imported and planted far ahead of their ecological time. The couple last returned to Reid in 1956. The cabins are an interesting part of Glacier Bay's history, and campers should refrain from using the wood for campfires.

There are several hiking opportunities from Reid Inlet, including an overland trek to Ptarmigan Creek (see chapter 10).

Johns Hopkins Inlet
Reid Inlet to Topeka Glacier
Distance: 9 miles
Paddling time: 4– 7 hours

Few places in Glacier Bay, maybe in all of Alaska, are as stunning as Johns Hopkins Inlet. Reaching this sheer-sided fiord, which features three tidewater glaciers near its head, is the unforgettable moment of many kayak trips into the park. Although it is only 9 miles from Reid Inlet to the gravel fan of Topeka Glacier (one of the few places to camp inside Johns Hopkins), the paddling is slow because kayakers have to weave through and negotiate the icebergs that continually flow out of the inlet.

From Reid Inlet paddle 2.5 miles along a steep and rocky shoreline to the northwest to reach Ptarmigan Creek and the gravel and stone beach on both sides of it. This is another popular and scenic camping area. At its east end there are more artifacts left over from Glacier Bay's mining era, including a wooden spoke wheel and the hull, or what's left of it, of a rowboat. Near the ruins is the mining road, now just a trail, that was carved up the valley toward the head of Ptarmigan Creek (see chapter 10).

The dominant point to the west is Lamplugh Point, which hides Lamplugh Glacier on the other side of it. North of it is another major point, and the pair mark the entrance of the 9-mile long Johns Hopkins Inlet. The ice will begin to gather in front of Ptarmigan Creek, and paddlers, in any kind of kayak, have to use extreme caution when weaving their way through. It takes but one good run-in with a large berg to crack the fiberglass bow of a kayak.

Tidal conditions, wind, and weather all play a role in how thick the icepack in Johns Hopkins will be. Generally paddlers have few problems crossing over to the north entrance of Johns Hopkins, where they are greeted with a view of Lamplugh Glacier. The .5-mile-wide body of ice will occasionally calve and is especially scenic in late afternoon, when the glacier and the icepack around it shimmer in the low angle of the sun.

Mouth of Johns Hopkins Inlet and the beginning of its icepack

Just west of the point that marks the north side of Johns Hopkins' entrance is the first camping spot in the inlet. The gravel shore that extends from the point provides good views of Lamplugh Glacier but none of the Johns Hopkins, unless you scramble high up the ridge behind the point. Good channels in the ice often follow the north shore as a point, 1.7 miles to the west, acts as a barrier in keeping the icepack away from the coast. The cove just east of the point is the next good camping area and even possesses some dryas-covered platforms. Again from the shore you can view Lamplugh but not Johns Hopkins Glacier.

For most people, the head of Johns Hopkins Inlet comes into sight when they scramble up the knob on this point or begin to paddle around it. Upon reaching either view point, most visitors are rendered speechless. Johns Hopkins Inlet sums up all the wonder of Glacier Bay. You don't paddle this section, you *experience* it, and once you've departed you'll never forget it.

Johns Hopkins Glacier, named after the university that sponsored an early expedition here, tumbles out of the peaks of the Fairweather Range to create a scene of rock, ice, and snow painted in grays, blues, and whites. Adjacent to it is Gilman Glacier, and up the inlet, Hoonah Glacier, both of whose snouts reach the water. Though Johns Hopkins Glacier appears to be less than 2 miles from the point, the face of the ice is actually 7 miles away. The view of the glacier is best in early morning during slack tide, when the water is calm and the sun lies behind you to the east.

Once around the point, paddling becomes much more difficult through the thick icepack along the north shore. Still kayakers often reach the gravel fan of Topeka Glacier as the runoff stream forces the ice away from the land. The spot offers some flat space for 2-person tents but is usually covered with snow until July. A view of Johns Hopkins Glacier is blocked by the knob just east of Toyatte Glacier. The outwash of Topeka Glacier, west of the stream flowing from it, provides an interesting 1.5-mile trek up the wall of the inlet, where you can come close to reaching the small ice mass.

Beyond the outwash, the ice is usually too thick for kayakers to travel without worrying about getting trapped when a rising tide closes the channels. Three-quarters of a mile east of the Topeka Glacier are what some call Chocolate Falls, as the dark brown streams make an impressive drop into the inlet. From the falls it is still 1.5 miles to the knob that blocks a clear view of the glacier from shore. Paddlers will find it difficult to reach the knob, and this deep in the inlet, equally hard to cross over to the south shore.

The south shore, beginning at the east side of Lamplugh Glacier, seems less open and harder to travel. Once around the glacier, you'll find that the ice becomes less congested. Paddlers sometimes thread their way to first Confusion Point, almost 2 miles to the west, and then to the dominant Jaw Point in another mile, where the inlet swings to the southwest. Once around Jaw Point you are greeted with good views of the head of Johns Hopkins Glacier. To continue farther along the south shore is difficult, and camping sites beyond Lamplugh Glacier are hard to find, existing only in the form of cramped spots in the few breaks of the steep shoreline.

Whenever camping in the inlet on the north or south shore, be prepared for the strong, cold winds that whip off the glaciers and down the waterway.

Tarr Inlet
West and East Shore to Russell Island
Distance: 24 miles
Paddling time: 2–3 days

Tarr Inlet sits cater-corner to Johns Hopkins Inlet. The head of it is another stunning scene of ice and snow and offers kayakers the rare opportunity to pitch their tent within view of two glaciers—Grand Pacific to the north and the smaller Margerie on the west side. U.S. Geological Survey (USGS) maps show Grand Pacific poised at the U.S.-Canada border, but the glacier is advancing and its face is almost a mile within Alaska; in 1986 it was only 100 yards from Margerie.

Paddlers coming from either Ptarmigan Creek or a campsite along Johns Hopkins Inlet's north shore have a 14- to 15-mile day ahead of them to reach a camping spot near the head of Tarr. Plan on a good 8 to 10 hours when heading into Tarr, as there will be ice to negotiate and often a northerly wind whipping down the inlet, especially on a clear day, making for choppy conditions and a harder paddle.

From the distinctive point that forms the north side of Johns Hopkins' entrance, the west shore of Tarr begins as a steep and rocky shoreline, with few if any places to land. This continues for 4 miles until you come to a small point with a stream and gravel beach following it. You pass another stream in the next 3 miles, before rounding the dominant knob in the middle of Tarr's west shore.

The cove formed by the knob lends itself to a scenic campsite but receives more than its share of wind and ice off the glaciers. The knob makes for a good scramble that gives you an excellent overview of the inlet. Beyond the knob, the ice gets considerably thicker as you approach Margerie Glacier 2 miles to the north. Just before you reach the glacier, the shoreline becomes a steep cliff that houses an impressive kittiwake colony.

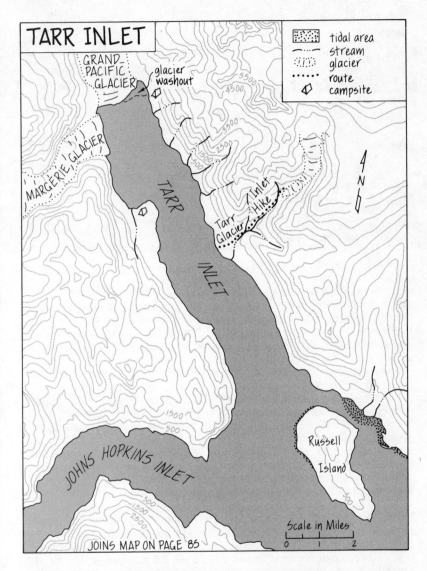

Margerie is an active glacier with a sheer face that measures almost 180 feet in height. Kayakers will marvel when it calves, rumbling and sending a series of swells through the upper portion of the inlet. By no means approach within a half-mile of the glacier. Depending on weather, tides, and wind, crossing over to the east shore of Tarr can be complicated and slow. The journey requires paddlers to thread their way through a maze of icebergs, which are generally fewer in number than in Johns Hopkins Inlet.

The face of the Grand Pacific appears streaked with gravel and stones and does not have the clean, sharp features of Margerie, its counterpart to the west. Nor is this glacier nearly so active. A half-mile south of the glacier,

Tarr Inlet from Reid Inlet (Ed Fogels photo)

along the east shore, is a small square point, covered sparingly with dryas, that provides limited space for 2-person tents and offers stunning views of the glaciers practically from your sleeping bag. Beyond this to the north is the gravel outwash of the Grand Pacific. Though flat, the outwash offers no protection against a possibly large calving from across the inlet that could quickly flood a campsite and sweep away any boat left untied. Streams are plentiful at the head of Tarr Inlet, but you might have to search for one not filled with glacial silt.

From the Grand Pacific, the east side of Tarr is for the next 5 miles mostly a steep, gravel shoreline broken up by numerous streams and extending to a large gravel outwash. Good lanes through the ice usually exist along the shoreline, and if you're heading south on a clear afternoon this portion will be an easy paddle, with the glacial wind whisking you along. The cove and point just north of the large delta feature a shingle beach with excellent campsites and good views of both glaciers. The delta itself is the start of an interesting trek if you want to view more glacial ice.

Tarr Inlet Glacier Hike (Distance: 2.5 miles one way; rating: moderate): The trek up this outwash and glacial valley is a very pleasant, half-day hike that turns into a trail of succession, beginning with alder and willow near the inlet and ending at the small arm of the glacier, 2.5 miles away. The hike is not difficult. You can follow the mesa-shaped lateral moraines on the north side of the outwash as they rise gradually for almost 2 miles.

Begin on the north side of the river, where the alder and bare rock banks meet, and hike northeast, aiming for the ridge of moraines. Many hikers prefer to trek along the moraines rather than tackle the boulders and soft banks of the river. Within a mile the dryas ends and you enter a bare but beautiful glacial valley where you can already see the tongue of ice to the east. When you hike across the stones, they look and even sound new; like cellophane wrappers rubbing together.

In a little more than a mile, three streams merge into the valley's main river, forcing you to ford the one to the north. In mid-summer this might involve hiking up the river a bit before you find an acceptable spot to cross. The large, flat moraines end in .5 mile, along with the easy hiking on their surfaces of gravel and patches of sand. At this point drop down to the river and continue working northeast through larger boulders, toward the ice, which is less than .75 mile away. In the final .25 mile, you have to scramble up the north side of the valley-turned-gorge to get closer to the ice before the major stream stops you. Fording the stream to reach the ice would be challenging by mid-summer, possibly requiring a hand line.

Russell Island

From the large outwash it is a 3.5-mile paddle to the entrance of Tarr Inlet, which is clearly marked by a wide point with a pair of distinctive rocky knobs. Paddlers heading north along Tarr's east shore will marvel at the scene before them when they round this point, as a panorama of both glaciers unfolds, deceitfully appearing to be only 2 or 3 miles away. Continuing south, in the next mile you round a second knob extending from the mainland (marked with the elevation of 510 feet on topographicals) and then slip into the channel formed by Russell Island's east side.

The 4-mile channel formed by Russell Island is usually a calm paddle, with little or no ice to worry about. Along the mainland the shoreline quickly turns into a 6-mile stretch of continuous shingle beach that provides excellent beach hiking and extends a good 2 miles beyond the south end of Russell Island. Campsites exist almost anywhere along this beach, with the Fairweather Range looming overhead, but the park service recommends backcountry travelers avoid it because of the high density of brown bears in the area.

Toward the north end of the channel is a wide gravel outwash along the mainland that turns into a valley leading back toward the west side of Mount Abdallah. By camping near the delta, you can spend a morning hiking up the valley, searching the mountainsides for goats. The east shore of Russell Island is rough, with few places to land and almost no suitable camping except a waterless spot at the north end. The south and southwest shores that face the entrance to Johns Hopkins Inlet are gently sloping.

Rendu and Queen Inlets
Russell Island to Gloomy Knob, including inlets
Distance: 36 miles
Paddling time: 3–4 days

You could paddle from the southeast end of Russell Island to the coves around Gloomy Knob in a day, but you would be bypassing two interesting inlets that only a small number of kayakers take the time to visit. Rendu and Queen inlets offer fiordlike scenery, a splendid collection of waterfalls, and two glaciers to explore, without the cruise-ship travelers and other kayakers that abound in Reid and Johns Hopkins inlets. The best bet is to plan on camping one night deep in each inlet to enjoy the solitude of being cut off from most other park users.

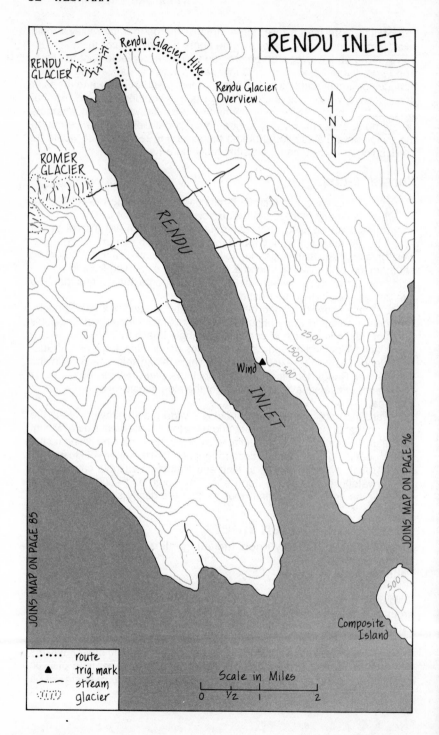

RENDU INLET

RENDU GLACIER

Rendu Glacier Hike

Rendu Glacier Overview

N

ROMER GLACIER

RENDU

Wind

INLET

JOINS MAP ON PAGE 85

JOINS MAP ON PAGE 96

Composite Island

····· route
▲ trig. mark
─·─ stream
〰〰 glacier

Scale in Miles
0   ½   1   2

Rendu Inlet

From the southeast end of Russell Island, it is a 3-mile paddle along the mainland to a pleasant little cove that lies just northwest of the entrance into Rendu Inlet. The cove features a nice strip of beach and a stream. It's also well protected, which is good because this stretch from Russell Island into Rendu can often be choppy, exposed to either northwest or south winds whipping through the bay. The paddle around the steep point into Rendu is especially hard on a windy day. Once you leave the cove, be prepared for 3 miles of steep shoreline, with no place to land.

When you reach Rendu Inlet, plan on spending 3 to 4 hours paddling the 8-mile inlet to its head near Rendu Glacier. The glacier has receded well onto land, so there should be little if any ice in the waterway. But there are waterfalls, and they have turned Rendu into a gallery of tumbling water. You will be treated to all kinds of waterfalls: braided ones, thundering drops that leap three or four times along the steep side of the inlet; others that appear out of nowhere, leaping 10 feet and then disappearing among the rocks again. At low tide you can paddle your boat along the shore and fill your water bottle, without getting more than your wrist wet.

After a 5-mile paddle up the west shore, you will come to a 2-mile stretch that contains three streams and their small gravel fans, the one to the north being the runoff from Romer Glacier. Each provides limited space for small tents on dryas-covered ground and offers a spectacular waterfall nearby. These campsites are unmatched on a sunny day in the middle of a long expedition. Nothing equals the invigorating feeling you get when standing under the falls with the ice-cold water hurling over you from head to toe. From the Romer Glacier stream, you go another good mile before reaching the extensive mud flats and runoff streams from Rendu Glacier.

The east side of the inlet is as steep as the west side and features just as many waterfalls. The falls are well worth the extra miles you go along the shoreline to view them; at times there will be a dozen waterfalls within 100 yards of each other. Some rough campsites lie 2.5 miles up the east side on a small gravel beach with protective rocks on both ends, just as you begin to round the wide point labeled on topographicals by the trig mark "Wind." The steep shoreline continues for another 4.5 miles until you reach a dominant knob on the east side across from Romer Glacier. There are places to beach a kayak and camp to the south, and the knob itself is a short scramble to good views of the glaciers to the north.

Rendu Glacier Hike (Distance: 2 miles round-trip; rating: moderate): Rendu Glacier is not visible from the gravel delta at the head of the inlet, and to approach it for a better view requires a mile hike over mounds of black ice. Begin at the center of the delta near the black ice between the two major meltwater streams from the glacier. By hiking almost due north, you reach a tongue of stagnant ice within .75 mile. Circle the ice to the east and climb up the moraine that parallels it. By following the moraine, you will swing toward the receding snout of Rendu Glacier. The black ice on the east side of the gravel delta has a considerable number of crevasses, and it's best to avoid traveling on it. As always, take extreme care when crossing such ice, as the gravel is loose and will quickly expose the slick ice underneath.

Rendu Glacier Overview (Distance: 4.5 miles round-trip; rating: moderate to difficult): This is a very scenic trek leading to fine view points of Rendu Inlet and its glaciers as well as much of the West Arm. Begin at the east corner of the gravel delta at the head of the inlet, on the east side of the main stream. Hike 1 mile in, climbing slightly along the gravel moraines. Within a mile you will come to a stream thundering out of the mountains into the meltwater of the glacier.

At this point turn east and scramble up the ridge. You'll encounter brush, but by staying near the mountain stream and the gorge it cuts, you will have a steeper climb with less alder. In a half-mile you come to a clearing and a flat area of dryas, a good camping spot if you're ambitious enough to haul your gear up there. Wander to the edge of the gorge to the north to view a spectacular waterfall that drops 20 or 30 feet straight down.

From the flat area, swing south and continue to ascend to a low point between two ridges. Within a half-mile you reach the pass, from where you can scramble up the ridge to the east. Follow this ridge to the first knob at 1,400 feet, which gives you superb views of the entire area. The ridge to the west can also be hiked but involves steeper sections and more challenging route-finding.

Queen Inlet

From the tip of the point that divides the two inlets, it is a 4-mile paddle up the west shore of Queen Inlet to the extensive tidal mud flats at its head. The only ice in Queen Inlet is what drifts in from the main bay, and there is little of that; Carroll Glacier is only visible as you enter the waterway. Up close from the water, the glacier is hidden behind hills of gravel and loose dirt.

The west shore of Queen, like much of Rendu, is an unbroken stretch of steep and rocky coastline. There is no place to land or camp until you reach the edge of the massive tidal flats, which now extend out to encircle Triangle Island. The only place to pitch a tent on either shore of upper Queen Inlet is north of the major stream and its gravel fan along the west shore. With Sentinel Peak looming behind you, this spot will keep you clear of the mud flats at low tide. A good hiking beach connects it to the massive gravel moraines, deposited by the retreating Carroll Glacier, that begin just a mile away.

The east side of the inlet is also steep, and mud flats make it difficult to find a suitable spot near the head of the inlet. Nor are there any camping opportunities along the shore to the south until you reach the stream and outwash that mark one side of the inlet's entrance.

Carroll Glacier Hike (Distance: 3.5 miles one way; rating: moderate to difficult): The easiest way to handle the terrain along this route is to begin with rubber boots or tennis shoes and then switch over to hiking boots once you reach the gravel hills in front of the glacier. From the northern corner of the inlet's large delta, follow the base of the ridge as it first heads east and then swings northeast, crossing many small streams along the way. Within 1.5 miles, small stones turn to boulders. You will be able to spot a main channel flowing between the ridge and the gravel moraines south of it. Cross the stream at first opportunity to be on its southeast side. In mid-

Waterfall in Rendu Inlet

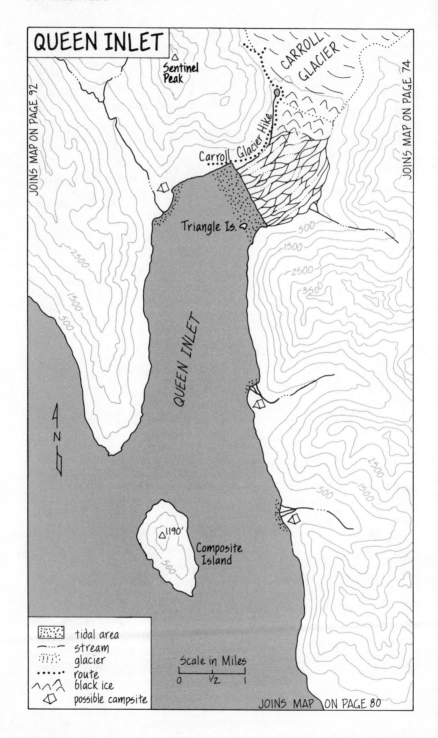

QUEEN INLET

Sentinel Peak

CARROLL GLACIER

JOINS MAP ON PAGE 92

JOINS MAP ON PAGE 74

Carroll Glacier Hike

Triangle Is.

500
1500
2500
2500
3500
1500

QUEEN INLET

2500

500
1500

N

△1190'
500

Composite Island

tidal area
stream
glacier
route
black ice
possible campsite

Scale in Miles

0   ½   1

JOINS MAP ON PAGE 80

summer you might have to search for a place where the stream is braided enough to easily ford.

The stream enters a steep gorge between the ridge and the gravel hills, and hikers are forced deeper into the loose mounds of dirt and ice because of the steep banks along the channel. Cut northeastward through the hills toward the glacier, working your way around each ravine and steep gully you come to. By staying near the stream, you will eventually (in a mile or so) come to a pair of glacial ponds that appear to be its main source. The glacier should also be visible at this point, as its edge lies .7 mile to the north.

Drop down and skirt the ponds and return to trekking across the hills, working your way toward the base of the ridge and following it to the edge of the glacier. For an overview of Carroll Glacier, hikers can scramble up either the ridge (which now forms the east slope of Sentinel Peak) or the high lateral moraine along its side.

Tidal Inlet
Queen Inlet to Tlingit Point, including Tidal Inlet
Distance: 27 miles
Paddling time: 2 days

The first gravel delta along the east side of Queen Inlet is 2 miles south of Triangle Island, at its head. The delta offers camping opportunities plus an interesting trek along the stream back to a deep gorge. The second delta is 1.25 miles farther south and across from the north end of Composite Island. More campsites on dryas-covered ledges can be found, and the stream leads east into a bowl that is surrounded on three sides by 1,500- to 2,000-foot granite walls. The bowl makes for both an impressive view from your campsite and an interesting scramble away from the shoreline.

The coast turns steep once you pass the gravel fan, and it remains that way for the next 3 miles until you reach the cove north of Gloomy Knob. Emptying into the cove is one of two large streams that flow out of Vivid Lake. The streams can be followed to that hidden body of water, though the brush gets thick at times. The cove is an extremely scenic campsite with a panorama, almost due west, of Mount Fairweather and the rest of the range. Strawberries abound in July; and nearby is the easiest route to the top of Gloomy Knob. (See "Gloomy Knob Hike" at the end of this section.)

It is a 2.5-mile paddle around Gloomy and an interesting trip either north or south, as on one side of you will be the Fairweather Range and on the other the steep, stone walls of Gloomy. Once around it you will come to the cove on the south side of the peak, where the other stream from Vivid Lake flows in. Plenty of campsites here, but no views of the mighty range to the west. Nearby is the southern route up Gloomy Knob (see "Gloomy Knob from the South" at the end of this section).

A small hammerhead peninsula separates the southern cove and stream of Vivid Lake from the entrance of Tidal Inlet, where good camping spots exist near the mouth on both sides. The stream that follows the east side of Black Cap Mountain is an interesting place to explore, with camping possibilities near the edges of its gravel fan. Across the inlet is a narrow and small spit of land that has good views and camping spots, but no water.

Fairweather Mountains from the West Arm

Once inside the entrance, the inlet narrows to .25 mile and becomes a scenic fiord extending 4 miles to the east, where on a clear day you can easily see the entire inlet. The walls of the inlet are sheer cliffs the first 2 miles and, like Rendu and Queen, a showcase for waterfalls. The north shore remains steep right back to the slopes of Pyramid Peak, which form the north side of the narrow valley at the head. If you plan to spend a night here, this is the only place to pitch a tent deep inside the inlet.

South of the entrance to Tidal Inlet, the shoreline continues to angle to the southeast, and within a mile you pass an islet that at low tide is connected to the mainland. The shingle beach here is the last spot to land for the next 6 miles because the shoreline becomes too steep farther on. This section is exposed to wind, from either the north or the south, which can make for choppy conditions along a coastline that often is nothing but towering granite walls. Waterfalls are impressive along this paddle, with several tumbling straight into the bay from 10 or 15 feet above. To the west you will enjoy a panorama of the Fairweather Range.

The first spots to land and camp are near the trig mark "Case," just before the shoreline curves east toward Muir Inlet. The small cove offers a pair of sandy beaches that make for fine campsites with superb views of Mount Fairweather to the northwest. From the cove you begin to round the large point that separates the East and West arms, and in 3 miles you'll reach Tlingit Point.

Gloomy Knob Hike (Distance: 2 miles one way; rating: moderate to difficult): The granite peak known as Gloomy Knob is 1,331 feet in elevation and usually a good place to view mountain goats early in the summer. Though it can be attempted from either the north or south sides, the northern route is a more gradual ascent, with less scrambling but more brush to contend with. Brush is both a blessing and a curse on this hike.

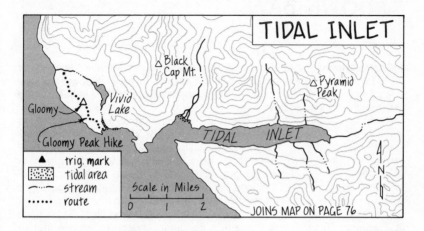

From the north, cross the river flowing out of Vivid Lake, which is usually a sneaker-soaking affair if you don't use your boat. Walk toward the mouth of the stream to begin climbing a prominent, gently sloping ridge. For the first quarter-mile, the alder and brush are thick, but quickly the grade becomes steeper and the alder thins out a little. Within a mile there is a small gully running north and south, with its west side composed of a steep, rocky wall. Hikers find it easier sometimes to hike in the gully and other times to continue through the brush on the ridge that parallels it to the southeast. After a quarter-mile they come together, and you continue to climb along a moderate grade, viewing two prominent knobs to the south.

Of these, the eastern one is the higher, but you scramble up to the low point between them, using ledges along the western one first. From the low point you finish climbing up the steep face of the higher knob by using ledges and cracks that curve around its face to the top. Choose your route carefully when attempting this section. Once on top you will be awarded with good views of Hugh Miller Inlet to the southwest and the Fairweather Range to the northwest. Almost due south is the true summit of Gloomy, and from here you dip down off the knob and climb up the final peak. The views from Gloomy, as might be expected, are extensive, and on a clear day you will be able to see a large portion of the West Arm and, to the west, Mount Fairweather and Mount Lituya.

Expect snowfields throughout June and possibly into July. Many of them will be hard and will make for easy passage through what would otherwise be an alder-infested route.

Gloomy Knob from the South (Distance: 2 miles; rating: difficult): This route begins at the last small shingle beach to the west, where you climb .75 mile first along patches of dryas and then bare rock. Keep the main ridge to the left and look for an alder-filled gully to the west marked by a prominent spruce tree near it. Use the tree as a marker, and scramble up to the gully east of it. In this way you reach the main ridge to the summit. From here you'll climb two false summits before spotting the true one rising 80 feet above a slight depression. The best route to the top lies to the right, but no matter how you attack it, it's a scramble on all fours to the peak.

7 DOWN BAY

Most kayakers arrive at the park and paddle up bay into the West Arm or Muir Inlet. But every summer a handful choose to explore down bay, the area that stretches from Dundas Bay along Icy Strait to Geikie Inlet on the west shore of Glacier Bay.

They forego the tidewater glaciers, barren morainal debris and tangled arms of alder to paddle into Glacier Bay's other world—one that is characterized by lush spruce and hemlock forests, ageless muskeg and layers of moss that carpet rocks and rotting trunks in a spongy cushion of green. In this special section of the park, gray gives way to green, the sight of an occasional cruise ship to the lone fishing boat, the excitement of seeing a calving glacier to the enjoyment of finding solitude in an isolated cove.

The stretch from the west arm of Dundas Bay to the mouth of Geikie Inlet, a journey of 50 to 60 miles, is a less frequently chosen route, one that involves paddling a portion of Icy Strait. It also requires either arranging a plane or boat drop-off or extensive backtracking if you want to avoid the additional cost of a charter fee. Once out in Dundas Bay, only the most experienced kayakers continue heading west along the park's perimeter or cross Icy Strait to other waterways.

Down Bay is neither for paddlers new to bluewater kayaking nor for those who want to spend most of their time near the park's frozen monuments. But for many, this coast's isolation, abundant wildlife, excellent stream fishing and rugged maritime scenery are ample rewards for the extra time and expense needed to paddle it.

The route is described from Dundas Bay north to Geikie Inlet, where kayakers can continue their journey up bay toward the glaciated areas of the West Arm or cross over to the east shore to head south toward Bartlett Cove and a return to civilization (see chapter 6). Those contemplating the lengthy two- to three-week paddle from Dundas Bay to Tarr Inlet should consider beginning in the southern portions of the park. The dramatics of the tidewater glaciers, when experienced first, often overwhelm and prevent you from fully enjoying the subtle and unique beauty of areas down bay.

The kayak journey beyond Point Wimbledon, the west side of the entrance to Dundas Bay, into Taylor Bay is not included here because the trip is only for experienced kayakers who can deal with the strong tidal currents and heavy swirls found in North Inian Passage and Cross Sound. Between the Inian Islands and Point Wimbledon, currents can reach the velocity of 5 to 6 knots on outgoing tides and can produce dangerous rips and swirls in the channel. Both this passage and all of Cross Sound are exposed to the often stormy weather and strong western winds off the Gulf of Alaska. Paddlers who wish to visit Taylor Bay and the Brady Glacier, however, can do so by hiking from the West Arm of Dundas Bay.

The North Passage of Icy Strait also requires caution and a readiness to hole up until good paddling conditions exist. But the speed of the current isn't as great as in Cross Sound, and the only area of heavy tidal swirls that will be encountered by kayakers following the coastline is around Point Carolus. Paddlers that are dropped off in the west arm of Dundas often

Dundas Bay (Ed Fogels photo)

spend two to three days in the bay and another five to seven days to reach and explore Geikie Inlet. Those attempting the complete journey from Dundas to Tarr Inlet with a return paddle to Barlett Cove should plan on no less than three weeks.

Fishing in Dundas Bay

DUNDAS BAY

Dundas Bay has three sections: the main bay, the north arm, and the west arm. The terrain is rugged and heavily forested, and the shoreline is steep around most of the bay. Gentler terrain, however, can be found in the lowland area that separates West Arm from Taylor Bay, and on a broad alluvial fan at the mouth of Dundas River, which empties into the northeast corner. Black bears, especially in late spring and early summer, are a common sight around Dundas, so all kayakers should maintain clean camps and keep food well away from their tents.

The main bay is 3.5 miles wide at its entrance between Point Wimbledon to the west and Point Dundas to the east. It extends 8 miles to the northwest, ending at a scattering of islands near the north arm. Both sides of Dundas' entrance are composed of steep, rocky shorelines and cliffs, where even emergency landings would be difficult. The east side of the bay can get extremely rough during foul weather, since large waves roll in from Cross Sound and batter this shoreline. Along the shore, the first shingle beach doesn't appear until 3 miles past Point Dundas at a small point and shoreline shown on topographicals by the trig mark "Deed."

Beyond the point, the bay continues to curve sharply to the east in a wide cove and vast tidal-flat area that is marked in the back by the remains of Buck Harbeson's cabins. The prospector arrived in 1931 to work the mining claims that Doc Silvers had registered in the area and built two cabins at the point furthermost east in the cove. It is debatable how much gold Harbeson ever recovered, but he enjoyed the solitude the bay offered and lived there until 1964, when he died alone in his cabin.

To the north of "Deed" point, an open-water crossing of 1.5 miles, is the mouth of the Dundas River and the vast tidal area that surrounds it. The main channel of the river swings past a long peninsula to the west. It is possible to paddle to the southeast corner of its end, even at low tide, and land on a steep, sandy embankment. Just up the steep shore is the flat tombstone of Andrew Jackson, an Alaskan who was buried here in 1965.

It is easy to see why the man wanted to end his days at this spot. Looming over the mouth of Dundas River to the east is White Cap Mountain and a second peak along the ridge, and to the west are the grassy lowlands that in July are full of strawberries. The peninsula provides some of the best camping spots in the bay, as nearby grassy flats offer unlimited sites that are decorated with lupine flowers throughout much of the summer.

You can follow the sandy beach and grassy banks along the west side of Dundas River for a pleasant hike almost 2 miles up the river, before the pine forest and thick brush move in to end the easy stroll. The nearby mud flats at low tide are an excellent place to search out tracks. Anglers can find Dolly Varden and cutthroat trout early in the summer and pink and silver salmon spawning up river in late July and August. Keep in mind that with the salmon runs come black bears, who are usually quite active in the areas surrounding Dundas River.

By paddling almost due west for 2 miles, from the tip of the peninsula to the west side of the bay (skirting the extensive tidal flats along the way), you will arrive at the Dundas Bay cannery site, which was owned by a number of companies between 1898 and 1920. In the early 1900s, the cannery em-

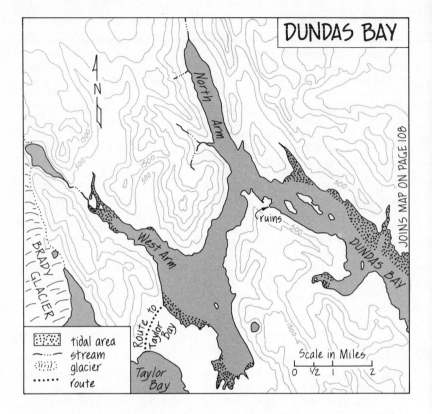

ployed more than 60 workers, half of them Chinese, and there were a re-
ported 40 homes stretched along the west shore north of it.

The ruins are impressive today, as an old, brick-wall boiler is still situated
high in the air on pilings. Nearby are the remaining timbers of a shipway and
a demolished barge half-hidden in the back of the cove. Scattered everywhere on the ground are gears, cables, wheels, large spikes, and other bits
and pieces of iron bleeding rust into the sand. There is also an old cabin,
built on pilings, that at one time or another in recent years housed park service rangers, researchers, and weary kayakers who sought its cover from
the rain. Inside is a guestbook for those who spend a night at the "Dundas
Bay Hilton."

A mile north of the ruins is a scenic bay that curves almost 2 miles to the
west and features extensive mud flats at its head. A steep-sided island sits
in the mouth of the bay, making it a haven of calm water for an occasional
paddler or commercial fisherman. The best camping sites lie a mile in, along
the south shore, where the grassy flats are level enough to allow you to
land. Toward the back of the bay paddlers will encounter mud flats that
make landing difficult.

Heading north you begin to paddle among the more than a dozen islands
and islets that lie in the northern portion of the main bay. Campsites exist

Dundas Bay (Ed Fogels photo)

only as grassy flats scattered here and there; they lack water but also few worries about bears. The southwest end of the second largest island is one of the better spots. The north end of this island lies near a cove along the west side of the bay, where there are more ruins, mostly logs, pilings, and cables, of Dundas Bay's commercial days. The mud flats in the cove make it a less than desirable campsite but an excellent place to view and photograph tracks of bears and other animals.

North Arm

To many, Dundas Bay's north arm is its most scenic avenue, as it extends 3.5 miles through a steep-sided fiord. The breathtaking beauty of this arm lies in the north end, where it appears to be almost totally enclosed by sheer walls of granite, so steep and straight that at times you strain your neck looking for the top of them.

The bay is less than a half-mile wide in most places; it offers calm conditions for paddlers and has a few choice campsites, all on the west side. The first is less than a mile up the arm, where a small river splits two ridges and empties into a shoreline that is sandy but without massive tidal areas. About a half-mile from the end, there is more sandy shoreline around the mouth of a stream, and grassy benches for tents where you can camp with spectacular views of the granite walls. A small river empties into the head of the arm and is surrounded by grassy flats, but here you must deal with wet terrain, muskeg, and the possibility of bears wandering down the valley.

Hiking in the arm is difficult, especially if you are not willing to crash through thick brush. The exceptions are a couple of avalanche chutes, where repeated snow slides have cleared a steep route up the ridge. The best scramble lies on the east side, halfway up the arm, where there is a low saddle in the ridge. Here the chute almost reaches the water and is stripped bare of anything but large pine trees. You can climb up to the peak on the south side of the saddle, a one-way trek of a mile or so, for an excellent overview of the rest of the bay and the Fairweather Mountains to the north.

West Arm

At its north end, the west arm of Dundas Bay is crowned with a limited view of Brady Glacier—just a slice of ice from the seat of a kayak. From here the arm stretches almost 6 miles south to the lowlands that form the mile-wide neck between Dundas and Taylor bays. To the east is the 2-mile-long channel that leads toward the main bay and the mouth of the north arm. Those who spend a night in the west arm usually choose to pitch their tent at the scenic north end, where a river empties into the arm after flowing through an area of grassy flats and shoreline. The spot offers anglers opportunities to wet their line and hikers a chance to trek inland along the west bank. At the head of the west arm, like everywhere else in Dundas Bay, the mud flats can be extensive at low tide.

Brady Glacier Hikes: There are three natural routes you can take to view Brady Glacier, which fills the head of Taylor Bay just to the west. All involve cross-country trekking through brush and require at least a half-day to complete. The best route begins where the west arm opens up to its widest point. To the east is the channel leading to the rest of Dundas Bay, to the west the start of the strip of lowlands that form the barrier to Taylor Bay. By cutting west across the northern end of the lowlands, you skirt the ends of two ridges and work your way through brush and spruce forests for almost 2 miles to Taylor Bay. Once you reach the other bay, you round a point and begin trekking over the vast tidal flats to the glacier. It is 2.5 miles across the steep flats to reach Brady Glacier, whose face is 2 miles wide and 400 feet high and appears ragged with dark streaks of mineral deposits. If you plan on trekking close to the ice, set aside a full day for the side trip.

At the head of the west arm, you can take another route by scrambling up the ridge on the west side of the river and making a direct ascent toward the knob of 2,133 feet. You have to hack your way up almost .75 mile, definitely a challenging trek, before you reach clear areas and superb overall views of the glacier. The third route begins by paddling toward the south end of the arm and reaching the largest of three coves there at high tide. The terrain is considerably wetter here, consisting in many places of stretches of muskeg and scattered ponds, but the narrow neck of land is only a .8-mile trek to Taylor Bay and views of the glacier to the north. Tidal flats now extend this far south, but a hike across them to the ice would be a one-way walk of 4 miles.

Icy Strait
From Point Dundas to Ripple Cove
Distance: 17 miles
Paddling time: 2 days

Paddling Icy Strait can be a challenging journey through fast currents and choppy conditions, or at times an easy day slicing through a smooth surface in what appears as nothing more than a large placid lake. Wind, weather, and tidal currents play an important role here, and kayakers should wait until the conditions are suitable for their level of skill on open water. Attempt to paddle with the tide as much as possible, and by all means hit the area surrounding Point Carolus at slack tide, preferably high slack. Stay within a couple hundred yards of shore, and during choppy conditions keep in mind that the calmest water is the thin strip that lies between the shoreline and the almost continuous bed of kelp found in Icy Strait.

Just north of Point Dundas is a small, curved point (indicated on topographicals by the trig mark "Gard") with a shingle beach on each side that would serve as a last-minute place to sit out any sudden rough conditions. The next 2 miles of shoreline, from the point until you have curved well into Icy Strait, is rocky and steep, offering no easy place to land even in calm weather.

This steep section is finally broken up by a pair of streams emptying into the strait along a shingle shoreline. The first provides an exceptionally scenic campsite, especially on a clear evening as the white-peaked mountains on Chichagof Island are illuminated in oranges and reds from the setting sun. From the stream it is a 2.5-mile paddle into the largest cove on this stretch of Icy Strait.

Emptying into the cove is a small river that during high tide can be paddled up to the lake where it originates. Camping along the river is not the best, as the views are limited, the terrain and surrounding mud flats often wet, and the bugs thick at times, owing to a lack of wind. Anglers will find Dolly Varden and cutthroat trout, but it is best to stay just outside the cove along the point that forms its east entrance, and paddle in with your fishing pole in hand. An even better campsite is a mile farther east, where a small stream empties into the strait from a sandy shore.

As you round the steep, rocky shoreline for the next 2 miles, the islands off Point Carolus eventually come into view to the east. Just before you reach this corner of Glacier Bay, you pass the wide mouth of a river and the sandy spit that forms its west side. This is another scenic spot to spend an evening, especially if you have to wait for better tidal conditions at Point Carolus. Campers will rejoice in the wide, sandy beach, large supply of driftwood, and ripe strawberries in July. Hikers and anglers can trek up the tidal flats for a spell and try to catch something fresh for dinner.

Point Carolus

A mile northeast from the river is Point Carolus: *When paddling around it use extreme caution.* Time your departure so that you reach the point at slack tide, slack high tide being preferable. The point itself is a low shoreline of gravel and boulders, with rocks and an extensive reef offshore. Large

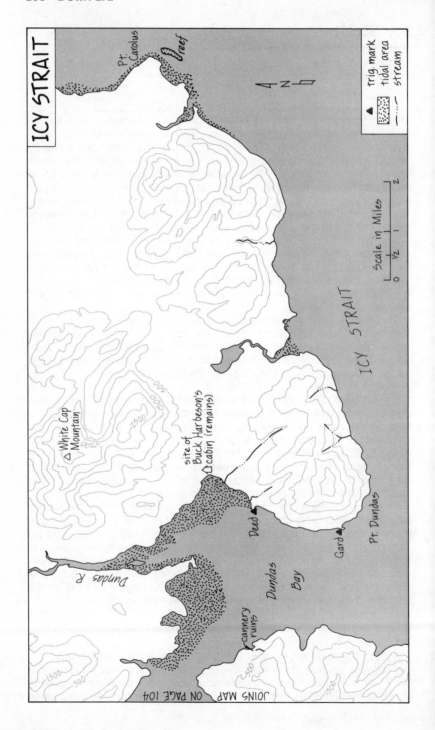

ICY STRAIT

Pt. Carolus

Reef

White Cap Mountain

site of
Buck Harbeson's
cabin (remains)

Dundas R.

Deed

Gard

Pt. Dundas

Dundas Bay

cannery ruins

ICY STRAIT

trig. mark
tidal area
stream

Scale in Miles
0   ½   1   2

JOINS MAP ON PAGE 104

vessels stay at least a mile away from the point, but kayakers should slip between the reef and the mainland, keeping an alert eye out for submerged rocks. Even at slack tide, small whirlpools will appear; if you slip into one, use strong strokes to continue paddling straight out.

Once around the point, you will have an immense view up bay as well as to the east, where you will be able to see Point Gustavus on a clear day. Within a mile of heading north you'll come to a shallow cove with a shingle beach 1.5 miles long; a good place to land after a hard paddle around Point Carolus.

There is little water on this beach, and none along the next 2.5 miles to Ripple Cove. This large cove features a long shingle beach and plenty of camping spots, but again, water is scarce most of the summer. In fact, running streams are hard to find in the first 8 miles north from Point Carolus, until you reach the cove on the north side of Rush Point. Whether you are going up bay or down, when paddling this stretch, depart from your previous camp with a good supply of water in your boat.

Berg Bay
From Rush Point to Fingers Bay
Distance: 17 miles
Paddling time: 1–2 days

Once beyond Rush Point, you enter the Sitakaday Narrows, a waterway of strong tidal currents that at times can reach 6 knots. For the next 5 miles, until you reach the first channel into Berg Bay, either paddle near slack tide or take an extended break along shore. It doesn't make much sense to battle rough currents when within a few hours conditions will improve.

Lars Island, at the entrance to Berg Bay, is for all practical purposes now connected to the rocky spit extending from the mainland. Neither offers any acceptable camping sites. Entrance into the bay is by way of the two channels north or south of Netland Island; kayakers should seriously consider at least dipping into the bay.

From the main bay, little of Berg Bay or its beauty is seen, but once you're beyond Netland Island, this body of water and the surrounding mountains unfold in all their splendor. The main stretch of water lies in front of you, while to the southwest the bay extends another 3.5 miles in an area that ranges from 6 to 14 feet in depth and is a haven for dungeness crabs. The west side of Netland makes for a scenic campsite, but the nearest water is a half-mile west of the island, where a good-sized stream empties into the upper part of the bay.

A much more scenic spot to pitch a tent is at the southwest corner of the head of the bay. The area is composed of tidal flats, but higher up there is a wide grassy bench cut across by a pair of crystal-clear streams. A backdrop of several 2,000- and 2,500-foot peaks looms to the west and southwest.

Coho and sockeye salmon spawn up these streams and in the river that flows into the northwest corner of Berg Bay; king salmon are known to at least visit the bay. Cutthroat trout and Dolly Varden can also be caught in the bay's fresh water. Anglers looking for a little adventure can hack their way .5 mile along the river to the northwest to reach a small lake (the first of several) where they can fish for trout.

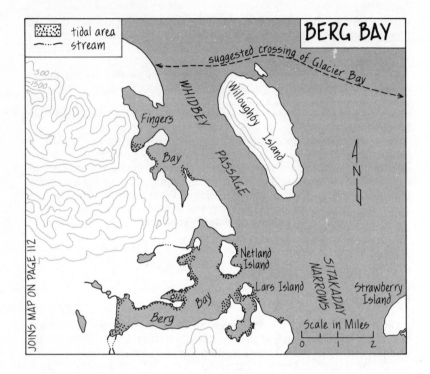

Fingers Bay

Dipping back into the main bay through the channel north of Netland Island, you round a 2-mile-wide head and quickly come into view of Fingers Bay. The 3-mile long but narrow bay is divided in half by its fingerlike peninsula whose shoreline is composed mostly of a steep gravel beach. Just north of the narrow base of the peninsula is a tidal area with a small island lying just offshore—the best, and maybe only, place to set up a tent on the north side. A more pleasant shingle beach and better campsites exist to the south, while to cross the peninsula between the two is a short trek through the brush.

Geikie Inlet
From Fingers Bay to the head of Geikie Inlet
Distance: 15– 21 miles
Paddling time: 1– 2 days

Lying between Fingers Bay and Geikie's Shag Cove is distinctive Marble Mountain. It is an 8-mile paddle to round but the trip is impressive as you move along beneath the steep sides of the mountain through Whidbey Passage, which is bounded on the east by Drake Island. Almost halfway along, some 3.5 miles from Fingers Bay, is a small cove with a gravel beach that provides easy landing. Looming directly overhead is a massive granite knob. No water here, but there is a grassy bench that makes for an unusual place to camp.

After rounding the northern tip of Marble Mountain, you enter Geikie Inlet, a beautiful body of water surrounded by peaks. The most prominent ones are to the west, where 4,100-foot Blackthorn Peak and 3,518-foot Contact Peak form what seems like a rock wall where the inlet abruptly ends. One of the best places to camp is 4 miles inside the inlet on the north shore near the lowland pass to Charpentier Inlet. The sand and gravel beach along this side can be hiked for miles, and usually you can gather enough driftwood for a campfire at night.

The route north to Charpentier Inlet is a gap between two ridges that was at one time a shortcut to the rest of Hugh Miller Inlet. This is no longer true, as the brush and alder are hideously thick and make carrying a boat through it impossible for all but the most determined souls.

From here Geikie Inlet extends another 6 miles until it ends at the ridge that forms Blackthorn and Contact peaks. To the south is the small cove where Wood Creek empties into the inlet. Although many maps show the creek as a wide river, it is not possible to paddle it into Wood Lake, the long lake located .5 mile south of Geikie. The bush is thick on both sides of the creek, and paddlers would have to haul their boat over to reach the scenic 2.6-mile long Wood Lake.

To the north at the head of Geikie Inlet is a large cove with extensive tidal flats, formed by the runoff of Geikie Glacier, and a classic glacial valley beyond. Good campsites abound on the edge of the flats and along the thicket of alder that is taking root quickly on both sides of the river.

Geikie Glacier Hike (Distance: 3 to 4 miles one way; rating: mild): The north tongue of Geikie Glacier had receded about 1.3 miles from the position shown on the 1948 USGS topographical and is now completely out of sight from the inlet. Those content with merely a good view of it can undertake an easy hike along the northeast side of its outwash stream. Beginning from the north corner of the tidal flats along the gravel beach, you can either pick your way for a half-mile through a band of emerging alder or drop down to closely follow the riverbank.

Expect snow to be lingering, even at this low level, well into early summer. The snowfields can be extensive; and you might spend as much as 1.5 miles tramping across them. Gradually, however, they give way to the gravel floor of the valley which is bordered by moraines and scattered with isolated colonies of dryas and alder. The walking here is easy as you follow a straight course for the distinctive knob that forms the north end of the Blackthorn Peak Ridge. Once near the corner, you'll see the terminus of the glacier, a snout dark with mineral deposits and sloping gently toward the ground. At this point it is possible to scramble up the loose gravel moraines along the valley for an excellent overview. The distance to the first good views along the northeast bank of the river is 3 miles.

Those intent on reaching the glacier, with the possibility of continuing on it, need to take to the southwest side of the river. Alder has also formed a .5-mile thicket on this side as well, but beyond it the valley is clear and makes for easy trekking. Near the knob that forms the north end of the Blackthorn Peak Ridge there are some interesting glacial pools that have to

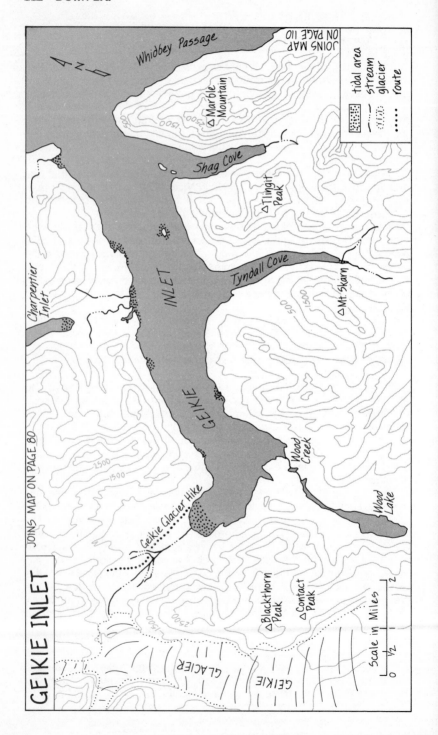

GEIKIE INLET

JOINS MAP ON PAGE 80

JOINS MAP ON PAGE 110

Whidbey Passage

△Marble Mountain

Shag Cove

△Tlingit Peak

Tyndall Cove

△Mt. Skarn

Charpentier Inlet

INLET

GEIKIE

Wood Creek

Wood Lake

Geikie Glacier Hike

△Blackthorn Peak

△Contact Peak

GEIKIE GLACIER

tidal area
stream
glacier
route

Scale in Miles

0   ½   1   2

Hiking up outwash to Geikie Glacier

be skirted around. The knob itself makes for a challenging scramble to a fine overview of the ice river. The trek to the snout of Geikie Glacier is a one-way hike of 4 miles.

Shag Cove

Marble Mountain becomes more stark and beautiful as you follow it up bay and finally round its northern tip into Shag Cove, a 2-mile arm of Geikie Inlet. The stark beauty persists on the east side of this scenic cove, where 3,000-foot rock walls drop straight into the water. Until early July, tongues of snow will spill out among the large boulders of the shoreline. So steep and barren is Shag Cove's east side that kayakers from the seat of their boats can often spot mountain goats scurrying along the ledges.

The west shore of the cove is well forested and steep in many places, but you can easily hike almost the entire east side, or camp at the head, where a stream empties into Shag. The only other possible campsites in the vicinity are on the waterless islands near the entrance.

Tyndall Cove

Two miles southwest of Shag is Geikie's second fingerlike cove. Tyndall Cove is 3 miles long and .75 mile wide at its mouth. From there it tapers down to the narrow tidal flats and two streams at its head, the only possible place deep within the cove to pitch a tent. The rest resembles a well-forested but steep fiord, with 2,500-foot Mount Skarn bordering its west side and 3,274-foot Tlingit Peak on its east.

PART III

GLACIER BAY ON FOOT

8 BARTLETT COVE

(See map page 54)

Breakfast No. 8 of the 1986 expedition: granola with honey and powdered milk, instant coffee, dried apples... and black bear. It was a large adult, maybe 300 pounds or so, and it wandered out of the woods as Bill and I huddled under our kitchen tarp, trying to keep our heads dry while waiting for our granola to get soggy.

In the early morning downpour, this bear didn't seem to be headed for any particular place. It emerged from the spruce trees 60 yards to our left, wandered toward the beach, and then casually stopped when something caught its interest. The bear scratched the ground, dug a little, sniffed a little, dug a little more, and then gave up.

The second time the huge animal paused, it was in the middle of a clump of grass to its liking. The bear promptly sat down and began eating its breakfast, which made us postpone ours. We took a couple of steps in its direction and with a telephoto lens closed the gap between us. Through the steady rain we could see the grass hanging out of the bear's mouth, and we watched and wondered in amazement.

The morning began in damp sleeping bags, with the prospect of our second straight day of rain. But low spirits immediately soared with the passing of the bear. It was our twenty-eighth sighting in eight days, and each encounter was vividly framed in our minds. Each had been an opportunity to study the animal rather than just watch it run off into the woods. Each had been a rare treat in the wilderness enjoyed by two people who spend most of their days in the city.

We watched it for 10 or 15 minutes until the bear grew tired of the grass and wandered off in search of something else. Elated by the experience we returned to our kitchen tarp, where our excitement suddenly tapered off; our granola had gone beyond the desired sogginess. Breakfast No. 8 was a bowl of solid mush.

Bartlett Cove Forest Trail
Distance: 1 mile
Rating: Easy

The only maintained trails in Glacier Bay National Park are two paths in Bartlett Cove, one of which is the mile-long Forest Trail. The path begins on the east side of the lodge and heads south a half-mile through a pond-studded spruce and hemlock forest where every log and rock seems to be carpeted with a thick blanket of moss. Just before reaching the campground, the trail descends toward the beach, swings to the north, and returns to the lodge, passing the dock and visitor-information center at the end.

There are naturalist-led hikes along the trail daily during the summer that unfold the story of succession in Glacier Bay as well as identify local flora. Check the Naturalist Activities Board in the lodge for times and gathering places for the hikes.

Bartlett River Trail
Distance: 3 miles round-trip
Rating: Easy

This 1.5-mile trail departs from the road to Gustavus and ends at the Bartlett River estuary where ducks, geese, great blue herons, and various other water birds concentrate in large numbers during migration. In late July and August chum salmon struggle upstream by the thousands to spawn. Anglers also do well for cutthroat trout and Dolly Varden, while visitors content just to listen and watch might view black bears, coyotes, or porcupines.

From the lodge, head east on the road to Gustavus, passing the side road to the park headquarters on the lagoon and continuing another 50 yards uphill. Here a sign marks the trail head, from which a path descends near the lagoon and follows it for a short distance. From the lagoon, the trail reenters the spruce forest and soon passes the old side trail to Bartlett

Black bear in the north arm of Dundas Bay

Lake, .5 mile from the road. The old trail is very difficult to find and almost impossible to follow, as it has all but eroded away. The main trail continues in the forest and after crossing a small stream emerges onto the open area around Bartlett River.

The round-trip is a pleasant 2- to 3-hour walk and another destination of naturalist-led hikes mentioned earlier.

Point Gustavus Traverse
Bartlett Cove to Salmon River
Distance: 12 miles
Rating: Moderate

The hike along the shoreline south of Bartlett Cove is a pleasant walk on a beach of sand, shingle, rock, and sometimes mud flats, with a handful of streams to ford along the way. The highlight is Point Gustavus, an excellent place to camp and one from which scenic views of Icy Strait are enjoyed along with possible sightings of seals, orcas, eagles, and black bears. Those planning to hike into Gustavus should keep in mind that the Salmon River at the end is a deep channel that has to be forded at low tide; otherwise it is a chilly swim across. Plan on 6 to 8 hours for the one-way hike to Gustavus.

Beginning at the NPS dock, follow the path toward the park campground. The trail passes it and continues for .25 mile until fading out in the tall grass. Cut toward the rocky shoreline of the beach and in 1.6 miles or so you will reach your first cove and immediately begin rounding into a larger one. The shoreline remains rocky for the first 4 miles and then switches to a shingle beach broken up by short stretches of sand.

Point Gustavus is reached in 6 miles; here you will be able to view Icy Strait both to the east and to the west. This is the best spot for overnight camping as the beach is sandy, the driftwood plentiful, and the views—which include Point Carolus across Glacier Bay—are incredible. Point Gustavus is one of the best places to consistently sight the tall black and white fins of orcas cutting through the waves of Icy Strait.

Eastward the shoreline remains sandy and provides excellent beach-combing opportunities. Everything from the smallest marine animals to entire trees, roots and all, are blown or washed ashore. A small stream is encountered 2 miles from Point Gustavus, at which point old pilings of fish traps appear offshore. From here to Salmon River the mud flats are extensive during low tide, and gradually the thick spruce forest gives way to grassy open areas.

The next stream, at 4.5 miles from Point Gustavus, marks where mud flats replace sandy beach. One-half mile farther is another stream; here the Gustavus pier comes into view to the east. The Salmon River, a 12-mile hike from Bartlett Cove, is a slow-moving waterway at low tide. Even then, hikers fording the river should expect to get their shorts and shirt wet.

From the Salmon River, hike another .25 mile to intersect the road leading to the pier. Head north on the road to reach the main road back to Bartlett Cove. If you hit the Salmon River at high tide and attempt to follow the west bank to the main road, you will end up trespassing through private property.

9 EAST ARM

(See maps pages 71, 74)

Not everybody who ventures into Glacier Bay's backcountry is a kayaker. Campers, hikers, and others who want nothing to do with those strange-looking boats can still journey up bay and explore the park beyond its shorelines.

Hiking from Bartlett Cove north is unfeasible and rarely, if ever, done. Backpackers set on exploring Muir Inlet on foot use the park's tour boat and begin their treks from the various drop-off points in the arm. Beginning in 1987, the park service may rotate the sites if heavy summer use alters the surrounding terrain. Visitors who want to hike only portions of the park should double-check the current drop-off sites while planning their trips.

The long-time traditional sites are Muir Point, Wolf Point, and Riggs Glacier, though things may change in the future. At Muir Point the hiking is limited, as the 7-mile trek along the south side of the narrows into Adams Inlet is a difficult one that includes almost continuous alder and steep sections of shoreline.

But Wolf Point is a different story because it offers hikers the most opportunities to wander through Glacier Bay's incredible landscape. Several treks are possible from this drop-off point, including multi-day trips south into Wachusett Inlet and north to upper Muir Inlet where you travel along ice remnants of stagnant glaciers. Several short hikes are also possible from the Riggs Glacier drop-off, but extended trips out of the area in upper Muir Inlet require special equipment and expertise in glacier travel.

WOLF COVE

Campers and hikers are dropped off on the southern tip of Wolf Point, and from there they scramble along a steep shoreline for 100 yards into Wolf Cove, a beautiful spot to pitch a tent for a couple days. The icebergs no longer clutter the shoreline like they once did, and alder is moving in rapidly, but the cove still harbors large flats of dryas that make excellent campsites. The view is scenic all around, with The Nunatak rising due east and distinctive Mount Case and Mount Wright to the south. The beach offers pleasant walking for a half-mile to the outlet from the glacial lake in front of the Muir Remnant. This river is usually too swollen and deep for most hikers to ford.

White Thunder Ridge

Distance: 2.5 miles to 1,196 feet
4 miles to 1,900 feet
Rating: Difficult

White Thunder Ridge, which rises north of Wolf Cove, picked up its name at a time when it overlooked three glaciers—McBride, Riggs, and Muir—that were almost continuously calving (hence white thunder). Today, Riggs is considerably quieter, and Muir has retreated out of sight from the crest of the ridge. The ridge itself is changing, as alder is slowly choking what was once a clear hiking route to fine overviews of upper Muir Inlet.

The hike up to the 700-foot point is still quite clear and easy to follow, but the trek to the next high point of 1,196 feet requires bucking alder for .8 to 1 mile. Less alder is encountered on the final leg to the high point of 1,900 feet, where the views are magnificent. In the end, hikers have to judge for themselves if they are willing to buck almost a mile of thick alder to reach the upper portions of White Thunder Ridge.

Begin at the first small stream you encounter, walking west from Wolf Point toward the cove. The stream drains a low gap in White Thunder Ridge; to the south is a distinct 300-foot knob. Follow the stream until you reach its headwaters—a series of small ponds. To the northwest will be the 700-foot high point, the best route is the easy-to-spot, dryas-covered ledges along its west side. It is 1.25 miles from the cove to the high point, where McBride Glacier and parts of Riggs, along with much of upper Muir Inlet, can be viewed.

To continue on to the main ridge, descend from the knob, but to avoid deep gullies stay to the right as you head north. Alder will be encountered as soon as you begin your descent. You will continue to fight it as you work your way up and over two small rises, before approaching a steep climb for the next high point. Be certain to stay east of the rock face that can be seen from the knob and through the alder. The easiest route is to swing slightly west before curving back east to skirt the bare rock. Once past the rock face, it is a good .5-mile climb through the alder to reach 1,196 feet (trig mark "Nomo").

The route to the knob at 1,900 feet begins by departing the 1,196-foot knob in a northwest direction and first descending almost 200 feet. You'll run into more alder at the low point before you begin scrambling toward the next high point, following an ascent that swings from northwest to almost due north in the final .5 mile. It is a 1.5-mile hike from the 1,196-foot knob to the high point of 1,900 feet, where snow normally persists through the first two weeks of July.

White Thunder Ridge tops off at 1,900 feet and from here the views are stunning. It is possible for experienced hikers to head west and descend to the Muir Remnant. Few exit this way, however, for this leg is a steep drop on often loose or rotten rock to the ice below and from there you still have to circle back to Wolf Cove, making for a long day.

Wolf Cove to Upper Muir Inlet

Distance: 4.5 miles to 800-foot knob
6.7 miles to Upper Muir Inlet
Rating: Moderate to difficult

From Wolf Cove there are several routes to Muir Remnant, the large ice field that the retreating glacier left behind between White Thunder and Minnesota ridges. Alder will be encountered on all of them but you buck less of it by following the river that serves as an outlet for the glacial lake in front of the stagnant ice.

Begin by hiking south along Wolf Cove and fording the first stream you approach. This is a knee-deep crossing but usually manageable throughout the summer without having to search for a braided section upstream. A half-mile south along the cove you reach the main river which is wide and deep, even in early summer, and too swift for most hikers to ford and continue south.

Follow the river's north bank, and within .25 mile you'll find the first of many fine examples of interstadial stumps and logs, some still standing and embedded in the ground. The banks are low and clear of thick brush for the first mile, before the river swings toward high cliffs of loose gravel with alder rooted on top. During early summer or periods of low water, it is often possible to continue along the base of the cliffs for another half-mile or so until you are within sight of the thunderous waterfall. At other times you might be forced up into the cliffs as the river's swift current has eroded the bank.

The waterfall is impressive with its water exploding through a rock gorge. At this point hike up the high banks and around the falls through the alder for .3 mile—a small price for the sight that will soon greet you. Emerging from the alder, you arrive at a large lake at the base of the stagnant ice field, with towering Minnesota Ridge to the southwest and White Thunder Ridge to the northeast.

Free from the entangling arms of alder, you drop down to the lakeshore and head northwest toward the ice. It is 1.4 miles of easy walking along the shoreline to the southern tip of the ice field. Be careful of approaching and ascending the ice, and never enter the ice caves found near the southern end. But hiking on the remnant itself is easy and pleasant along a rough surface covered with gravel and rocks in many sections.

By staying on the eastern half you can walk straight back to the prominent 800-foot knob emerging from the middle of it. It is 1 mile to the base of this hill and another .3 mile up to its high point for an overview of this incredible left-over piece of glacier. From the top of the knob you can clearly view the route west through Glacier Pass and beyond to the Burroughs Remnant, while to the north is the route to the shoreline of upper Muir Inlet. Just north of the knob, the ice swings toward the base of White Thunder Ridge.

From the base of the knoll you can continue on the ice field to the north-

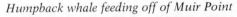

Humpback whale feeding off of Muir Point

west, where it leads to a pass between a pair of distinct 1,800-foot peaks. Eventually the ice field ends within the pass; you should refrain from entering ice caves found along this northern edge. From the pass you follow the boulders and rugged terrain to descend to the shoreline of upper Muir Inlet, where there are rough campsites for small 2-person tents (see chapter 5). It is 2.7 miles from the 800-foot knob to the shore of upper Muir Inlet.

An alternative to reaching or departing from the glacier remnant from Wolf Cove is hiking through the lowland area of ponds, though this involves considerably more alder and brush to battle. To reach the ice field, follow the first stream south of Wolf Point and stay with the channel to the west when it branches off. Hike 1.2 miles or so along this stream until the second gap in this low ridge appears. The scramble up will require bucking a lot of alder before you break out above the east shore of the glacial lake in front of the ice field. From the shore of the lake it is considerably easier to spot this gap.

Wolf Cove to Wachusett Inlet

Distance: 13–14 miles one way
Rating: Difficult

The trek from Wolf Cove to the northeast side of Wachusett Inlet is a long day of hiking, with the majority of it on remnant ice, an easy surface to hike on but a rough place to be if you want to quit early and set up camp. The overnight trip also involves climbing through Glacier Pass, which at 1,600 feet has snow well into August. Hikers contemplating the journey should be prepared for such conditions and plan on starting early in the morning to make full use of Alaska's long summer days.

Once you have reached the 800-foot knob on Muir Remnant (see previous hike), Glacier Pass is clearly visible to the west as a rounded "V" marking the north end of Minnesota Ridge. It is a good mile from the knob to the west side of the remnant ice where you begin to ascend to the pass, or a 5-mile journey from the shore of Wolf Cove. It takes almost a mile of climbing to reach the high point of the pass, with much of it on black ice (stagnant ice covered with a layer of gravel and rubble). Upon descending, swing to the south and follow the base of Minnesota Ridge to reach the remnant of Burroughs Glacier.

Hikers can head west or east on the vast ice field, but those who choose west should keep in mind that this end of Burroughs features a large meltwater river that has cut steeply into the north end of Bruce Hills. Either fording the river or climbing over Bruce Hills to reach the shore of Wachusett Inlet would be an immense challenge, if not an impossible one, for most backpackers.

Heading east, hike approximately 2.5 miles along the north side of Burroughs and then another .5 mile across the ice field to the east end of Bruce Hills. At the end of the hills is a high point rising 300 feet above the ice. Scramble to the top for a good overview of the glacial remnant, its two glacial lakes, as well as the terrain southward toward the inlet. Hikers heading toward the shoreline of the inlet, or on a day hike from there, will find it easiest to approach or depart the ice by climbing over the east end of this knob.

From the top of the knob descend into low moraines of gravel and loose dirt. To the east is remnant ice; to the southeast, the smaller of two glacial lakes that were seen from the knob. Work your way through the moraines and large boulders to skirt the lake around its west side, then ascend the low pass due south of it. The view north from this pass includes the Burroughs Remnant, Minnesota Ridge, and Bruce Hills. To the south you will see another low ridge and, in between, a stream that forms a thundering waterfall.

Hike south toward the next ridge, crossing the stream at the best possible place. Climb the low ridge and follow it to the west, keeping the stream in sight while swinging clear of the impressive falls. From this ridge a low area of ponds, scattered alder, and a small lake, which was left when Plateau Glacier retreated, appears to the southwest. The stream that forms the falls can easily be seen flowing through this glacial valley into the lake.

Descend to the east bank of the stream and, working around one small pond after another, follow it to the lake. Upon reaching the lake, you will most likely be forced to scale the high bluffs that border it to continue as there will be little shoreline to hike along. The last leg before getting to the inlet involves hiking a half-mile along the banks of the small river that flows from the lake's southwest corner to a gravel outwash on Wachusett's shoreline.

Kayakers who want to begin the hike from Wachusett should look for the second gravel fan on the north shore once they have passed the "island" some 2 miles from the mouth. The river is difficult to ford throughout most of the summer, so reaching the Burroughs Remnant requires beginning and staying along its east side despite the high banks that border it.

RIGGS GLACIER

To many people, Riggs Glacier is the most scenic mass of ice in Muir Inlet, and it has always been a popular drop-off point for the *Thunder Bay* tour boat. In the past, if you went to Riggs Glacier in the height of summer, you would find an interesting assortment of tents and kayaks stretched along the sand-and-gravel arm that forms the tidewater lagoon in front of the glacier's face. Because of its popularity, however, this area will probably be the first drop-off site that the park service discontinues.

As a camping spot, it is unsurpassed in the East Arm. Tents can be pitched with a view of Riggs' icy, blue snout, and many backpackers cherish the rare opportunity to drift off to sleep at night to the occasional rumble of falling ice. As a departure point for hikes, it offers a couple possibilities but overall is somewhat limited for those without expertise and the right equipment for glacial travel.

Campers who arrive without a kayak will find McConnell Ridge a major barrier to any trek south for close views of McBride Glacier. The shoreline between the two glaciers is a 3-mile trek along what is basically the west end of the ridge. The first 2 miles from Riggs Glacier is a very steep and rocky shoreline that is broken up by numerous deep gullies. All of this makes for an extremely challenging hike.

At the other end of the hiking spectrum is a leisurely mile-long loop following the inside of the lagoon during low tide for a closer view of Riggs' face. Leave an hour before low tide if possible, and be prepared for some muddy spots and a few streams on the tidal flats in the southeast corner of the lagoon.

McBride Glacier Remnant Hike

Distance: 2.9 miles one way
Rating: Moderate

Riggs Glacier has receded from its position on topographicals and now consists of the main trunk and a smaller arm that wraps around a rocky knoll just to the south. Next to this is a stretch of black ice and all that remains of McBride Glacier (shown on the USGS maps as sharing a common face with Riggs). The black ice is stable and can be followed for over 2 miles to good views of the main trunk of McBride and of Coleman Peak. Good hiking boots and caution are essential, for the gravel and rubble are loose and can quickly expose slick ice underneath.

Begin at the south side of the tidewater lagoon (best reached at low tide), which allows you to skirt along the water line. (At high tide you must scramble along the high rocky bluffs leading to the ice.) A braided stream empties into the lagoon from the southern arm of Riggs, but before you reach it begin climbing the gravel hills that lie between it and McConnell Ridge.

Numerous gullies and crevasses lie at the edge of the black ice near McConnell Ridge; it is best to swing north of them toward the small arm of Riggs. Within a half-mile, you will be able to walk on the edge of the arm, which is the best route to follow for the next mile or so. After the rubble mounds of the black ice, it is enjoyable to trek along this vast, white surface broken up only by an occasional deep blue crack or small pool of glacial water.

Within a mile, the arm curves to the north, and more black ice is encountered. To continue east toward the main trunk of McBride, swing toward McConnell Ridge and follow a distinctive parallel ridge of black ice. To the east McBride Glacier reappears as a clean ice field lying between McConnell and a steep ridge to the north. Keep a sharp eye out for mountain goats traveling along cracks and ledges on both ridges.

The black-ice ridge can be followed for almost another mile, until it merges into the side of McConnell Ridge. At this point you can see McBride Glacier pouring out of the mountains to the northeast, while straight ahead is Coleman Peak, an impressive 5,630-foot summit. Hikers toying with the idea of returning to their campsite via McConnell Ridge should be aware that the scramble up its steep northern side is an extremely challenging climb.

McConnell Ridge
Distance: 2 miles one way to 1,000 feet
Rating: Difficult

Another trek for those who are landlocked at Riggs Glacier or who simply want a day off from kayaking is to climb McConnell Ridge to the south, beginning from the sand-and-gravel spit. This is a challenging hike, where selecting the best route is important and scrambling on all fours is necessary at times on the way up. Sturdy hiking boots are a must, as is reasonably dry weather.

The hike begins with a scramble up the ridge from the spit. The easiest route is toward the west end of the ridge, near Muir Inlet. The .25-mile scramble is steep; you should choose your route carefully and be prepared for loose rock. On this slope alder becomes a blessing as a hand hold rather than a curse. The ridge levels off after .25 mile, and you encounter three knobs. The one farthest to the west gives a fine overview of Riggs Glacier and much of the upper arm of Muir Inlet.

Cut south across the ridge until you come to the edge of a deep gorge, whose opposite side appears as a steep, solid-rock wall. Follow the gorge east before descending into it and working your way to its eastern end, where you can view more of Riggs Glacier. Here is the best area for nontechnical climbers to ascend the ridge. It is a steep climb, so choose your route carefully. Eventually the way levels out considerably and you find yourself trekking southeast across numerous gravel hills toward the steep side of the knoll, marked on topographicals by the elevation of 920 feet.

After descending into another small gorge, work your way to the east side of the 920-foot knob, as the best routes to continue up the ridge are located here. This section is steep, but once the terrain levels out you follow a broken course over mounds of gravel, where alder has already taken a sure foothold.

At 1,000 feet you come to a small ravine, on whose opposite side there is an obvious route to the high point of 1,500 feet. Because of heavy snow, it would be difficult to hike this route before mid-July, perhaps even later, without ice axe and crampons. For those continuing, the climb becomes steep again to the outcropping at 2,000 feet, where it levels out slightly before resuming its ascent to 2,500 feet. McConnell Ridge tops off with a pair of rocky peaks at 3,000 feet, the first visible at 1,500 feet.

Above: Reid Glacier Overview hike. Below: The face of Reid Glacier in the West Arm

10 WEST ARM

(See map page 85)

Drop-off sites are limited in the West Arm and are due to be rotated if areas experience heavy use. Traditionally the park's tour boat has stopped at only two places in the upper reaches of the arm. One is Lamplugh Point, a small rocky landing on the east side of the glacier by the same name. It is used exclusively by kayakers, as the park service requests that no one camp within a mile of the site.

REID INLET

The other drop-off point in the West Arm is Reid Inlet, a scenic fiord crowned by the tidewater face of Reid Glacier and the mountains that form its steep sides. Good camping exists along much of the north shoreline of the inlet, though most people tend to congregate around the historical Ibach cabins near the mouth, where a small spit extends to the southeast. Both shorelines of Reid are composed of shingle beach; it is an easy 2-mile walk from the spit (near the mouth) to the corners of the glacier, where you get an unusual view of it. Avoid walking in front of the face on the mud flats during low tide, as the glacier is capable of calving without warning. Keep an eye out for, and avoid disturbing, tern nesting sites along the shoreline.

From the north shore of the inlet two hikes are possible: a climb to an overview of the glacier and a trek to Ptarmigan Creek that can easily be turned into an overnight trip.

Reid Glacier Overview

Distance: 2–4 miles round-trip
Rating: Moderate to difficult

A walk along the west shore that ends at the southwest corner of the inlet can easily be continued by ascending the side of the ridge that parallels Reid Glacier. It is a beautiful hike, with the first mile leading to a visible outcropping of only moderate difficulty. Constantly looming to one side is a massive wall of blue and white ice, while to the north is a sweeping overview of the inlet and the peaks surrounding Mount Abdallah. Every step improves your panorama of this stunning section of Glacier Bay.

Good hiking boots, not rubber boat boots, are needed for this hike, which not only covers a sharply angled ridge of loose rocks but also numerous meltwater streams that make for slippery footing in places. Also pack along a jacket or sweater, even on hot afternoons, as the glacier generates a cold wind that slaps you harder the higher you climb.

Begin by traversing up the ridge toward the distinct knoll visible above, a mile away and a climb to 700 feet. The trek will likely be a series of steps up and down as you work your way around lingering snowfields and over major runoffs. The alder has not reached jungle stage yet, and here it is helpful as a handhold. The way gets steeper the higher you climb, but scrambling on all fours is usually not necessary. The knob makes a superb place to lunch, offering a different view of the glacier's snout from above.

The next mile, leading to a knoll at 1,700 feet, is considerably harder, but the reward is a perch from which you can view more than 6 miles of Reid Glacier and much of the main bay around Russell Island. From the 700-foot outcropping continue to traverse the ridge while gradually ascending, working your way past a sheer rock bluff above. Once past the prominent high point at the bluff's south end, begin a steeper ascent to arrive at a major stream tumbling down from a saddle above.

The saddle may not be clearly visible from the steep side of the ridge, but the 1700-foot knoll to the southwest should be. Ascend to the saddle by following the stream. You will top out at what resembles (until late summer) a snow bowl, with a distinctive steep ridge surrounding it to the west. At this point the knob will be to the south, and it is easily reached by swinging through the saddle and climbing its back side.

The high point of the knob is marked by a tall and out-of-place alder tree, which somehow took root years before the others around it. The tree is bent, even cracked, as a result of the strong winds that whip down off the ice field. But it is still surviving; a credit to its resilience and the inevitability of change in Glacier Bay.

From the knob you can view an extensive section of Reid Glacier, but unfortunately not the Brady Icefield, which can only be seen by climbing the nearby ridges, which requires mountaineering equipment and expertise.

Reid Inlet to Ptarmigan Creek

Distance: 3 miles
Rating: Moderate to difficult

A trek from Reid Inlet to the beach where Ptarmigan Creek empties into the bay is a hike past the mining ruins and claims of the West Arm. It begins near Ibach's old cabins (see chapter 6), passes another cabin that is still being used today, and ends on an old mining road that leads to more ruins rotting away near Ptarmigan Creek. It is far easier to reach the alpine pass between the two shorelines from Ptarmigan Creek than from Reid Inlet. The route from the inlet begins with a steep scramble up a gully that is challenging with a pack on. Backpackers can extend the trek and turn it into an enjoyable overnight hike by including a side trip to the headwaters of Ptarmigan Creek.

From the west shore of Reid Inlet, the route begins with the steep, rocky gully just south of Ibach's cabins in what is unquestionably the hardest part of the trek. A stream rushes down the gully and empties into Reid Inlet. The first half is an easy climb, but the second half is considerably more difficult, especially during early summer, when the route is still under a blanket of snow.

An alternative route is to hike a third of the way up, then look for a dryas-covered break in the steep northern bank of the gully, which looks like an old stream bed leading up. Once on top of the bank, circle north, then northeast around the 1,300-foot hill above you by following a series of ledges that lead to the knob's east side. The abundant alder can be used as hand-holds. Snow lies on the ledge itself until early July.

Either route eventually leads you to the saddle between the knoll to the east and the towering ridge to the southwest. Just west of the knob is an alpine pond (the source of the stream thundering down the gully) and a continuation of the mountain pass. Views up here are immense, including all of Reid Inlet and much of the West Arm. The north side of the saddle is formed by a rocky ridge with two 1,500-foot high points that can be climbed either from their east end or through a low point in the middle between them.

The saddle appears as a vast snowfield through much of the summer and is marked in the middle by an 8-foot boulder. From here you begin to descend toward Ptarmigan Creek. A pair of sharp eyes might even spot the miner's cabin far below at 500 feet. From the western end of the saddle, you can see the valley where Ptarmigan Creek originates. This is a good point to drop packs and take a side trek to explore the headwaters of the river.

The easiest route is to skirt the ridge that forms the south side of the saddle to Reid Inlet. By staying high, you avoid the heavy brush and broken terrain near the creek itself. Eventually you swing southwest and then descend to the upper portions of Ptarmigan Creek, following it through an immense and impressive granite bowl, which is most likely snow-filled through the first half of the summer and loaded with blueberries in the second half. The walking is easy here as you scramble up one more steep drop to a fine view point beneath the hanging glacier that is the source of Ptarmigan Creek. Keep an eye out for mountain goats high above you on the mountainsides. To hike up the valley by following Ptarmigan Creek from its mouth is considerably harder, involving a lot more alder.

From the saddle you can also descend steadily toward Ptarmigan Creek through knee-high brush until you reach one of its branches. Willow and alder get progressively thicker as you follow this small stream to the northwest. The stream leads you to the high bank of Ptarmigan Creek, where you buck some alder and willow for a half-mile or so until you see a cabin. Hikers should not enter the cabin as it is occasionally used by miners who hold the claims to the active mines on the west side of Ptarmigan Creek.

Leading away from the old wooden shack is a definite trail that skirts the high bank of the creek and then drops into a gully where the mining road begins. This end of the road is easy to recognize as a well-used trail, but sections farther down are bushy or have become flooded by a small stream. Stay with the road (despite getting your boots wet) and you will emerge at the east end of the Ptarmigan Creek beach in the middle of more mining relics, including wooden wheels and the rotting hull of a dory.

11 CAPE FAIRWEATHER TRAVERSE

The trek around Cape Fairweather, from Lituya Bay to Dry Bay, is a rare wilderness adventure that combines the thunderous beauty of the Gulf of Alaska with the majestic charm of the Fairweather Range. This challenging trip is for experienced backpackers only. It is necessary to carefully plan the trip well in advance in order to avoid major mishaps in an area where one cannot afford to be careless. But for those with the experience and desire to endure heavy packs and sore legs from long days of bouldering, the 70-mile hike is an amazing adventure in a scenic wilderness unmatched elsewhere in Glacier Bay National Park, maybe even in the rest of Alaska.

This is a trek along a wilderness beach: Little climbing is involved and no elevation is gained. The hiking ranges from flat sections of soft sand where barefoot backpackers often trudge along with 60-pound packs, to miles of large boulders where the pounding on ankles and calves hopping from one rock to the next takes an immense toll on the legs. There are also places where hikers must dip away from the shoreline into the lush spruce forest and follow bear trails.

The walk is challenging, the packs heavy, but the scenery spectacular. On one side of you is the pounding surf of the Gulf of Alaska and an endless view of water interrupted only by occasional fishing boats and curious sea lions that follow passersby offshore. The constant crash of the surf is always with you on this trip, from the most dismal days to the softest nights, when the sound becomes a serenade that turns drowsiness into dreams.

The power of the waves, wind, and weather is evident throughout the hike, as the shoreline is littered with everything from crab pots, life rings, and mounds of fishing net, to whole trees (roots and all), Japanese plastic beer cases, and even occasional glass fishing floats that have made their way, undamaged, across the Pacific Ocean.

To the east lies the Fairweather Range, glorious peaks of snow and ice that separate the outer coast from the inner bay. This glaciated range is crowned by 15,300-foot Mount Fairweather, which is visible from almost any point along the hike if the weather is clear.

The distance between Lituya Bay and Dry Bay is less than 70 miles and in good weather could be hiked in six days by a hardy group. But six days of clear weather is a rarity, and most trekking parties budget eight to ten days for the journey. Weather changes quickly on the outer coast, where backpackers should be ready, and willing, to endure storm conditions and all-day rains. The wind during the summer is predominantly from the southwest and can be strong at times. For this reason most groups begin at Lituya Bay and hike northwest to Dry Bay to keep the prevailing wind at their back.

Within each party there should be a two-person raft and 200 to 400 feet of rope. Along the route, depending on the season and recent rainfall, there may be several rivers that require ferrying people and equipment across. Some will appear so turbulent that it is best to pull the boat back and forth by the use of two lines rather then depend on paddling. There is also the possibility, especially before mid-August, that a river may become so swol-

Floatplane in Dundas Bay

len it is impossible to cross even with a raft. Numerous groups have had to abort their trip because of high water and then seek out a fisherman to radio for a pick-up.

This is brown bear country, and park rangers recommend that backpackers not camp next to rivers and streams with salmon runs. Instead, they suggest setting up tents in the heavily forested zone that generally begins a couple hundred yards up the beach.

The other thing to keep in mind about this trek is that the outer coast is not only a true wilderness but a changing one. Neither a written guide nor even a topographical map can present an accurate picture of what to expect. Rivers change course, dry up, or suddenly emerge where before there was only a trickle of a stream. New bear trails suddenly appear while old ones are reclaimed in the undergrowth.

The purpose of this chapter is to help you decide whether you want to endure this trek and, if you do, to assist in the planning of it. Once out there, you are on your own in every sense of the word. Maps and written descriptions are valuable only as reference aids during the decision process. In the end, you choose routes and make important decisions based on what you know, what you see, and what your common sense and backcountry experience dictates.

Make no mistake about it—the outer coast is true wilderness—a remote, isolated, constantly changing wilderness. And because of that, maybe an everlasting one.

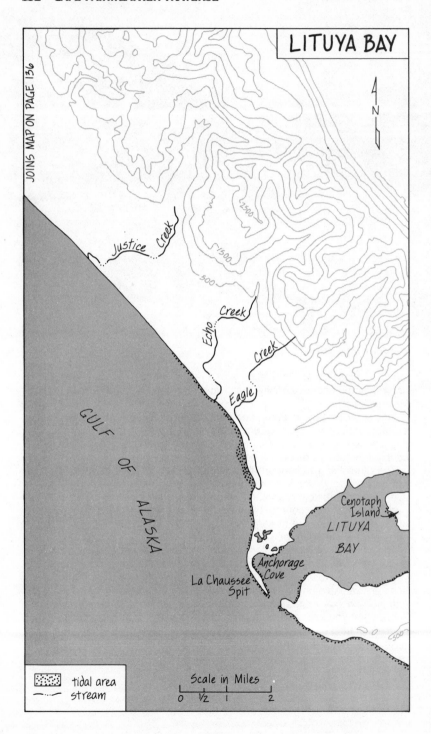

Lituya Bay
La Chaussee Spit to Justice Creek
Distance: 7–8 miles

To fly into Lituya Bay on a clear day is an unsurpassed way to start a wilderness hike. The 8-mile long bay runs from the sandy arm of La Chaussee Spit almost due east to a granite head wall anchored on each side by a pair of glaciers. Most hiking parties land in Anchorage Cove just east of La Chaussee Spit, from where they can view only the mountainous backdrop and Cenotaph Island, which lies in the middle of the bay. To view any more requires an extensive hike along the north shore.

La Chaussee Spit extends a mile from its base and is composed of huge boulders on the west side, sand on the east, and grass, thimbleberries, and small spruce trees in the middle. The views of the Fairweather Range and Cape Fairweather are stunning on a clear day from the spit. During midtides, the narrow entrance into Lituya Bay is a scene of wild water, which violently rocks any fishing boats trying to enter the bay. Indeed, it was here that 21 seamen from the crew of French explorer Jean LaPerouse perished on July 13, 1786, when two boats were drawn into the violent current and swamped.

Lituya Bay, one of the few protected bays on the outer coast, was a stop for most early-day explorers, particularly the Russians, and the site of a Tlingit Indian summer village. Cenotaph Island was also the home of the outer coast's only permanent resident. Sometime around 1915 Jim Huscroft rowed his boat into the bay, built a cabin on the island, and established a fox farm. He lived there, alone, for the next 24 years or so, departing only once a year to row to Juneau for supplies, the little mail he received, and a year's worth of newspapers that were kept for him by the local Elks Club. These he would read one a day, never cheating himself by skipping ahead among the year-old editions.

The spit is a scenic place to view the bay or spend a night, but beware that drinking water and a flat space to pitch a tent are limited and there is little protection from the wind and storms that sweep off the Gulf of Alaska. The campsites get better as you walk east along the north shore.

The trek north along the outer coast begins with bouldering. The rocks are large—for the first 2 miles to Eagle Creek you spend more time looking at the boulders than at the scenery around you. This is tough hiking as you step or leap from one rock to the next. Within .5 mile from the spit there are several large, rusty, and seemingly out-of-place pieces of mining equipment sitting on the shoreline between the huge rocks.

An alternative for this stretch is to move inland, 20 to 30 yards within the spruce trees, and search out bear trails or an old mining road to follow. The road runs from Anchorage Cove to Justice Creek but is difficult if not impossible to locate in many sections as it is overrun with brush, especially in mid- to late summer. The best stretch is along a small lake .5 mile north of the spit, where you can still see a pair of ruts along with an old truck and jeep sitting nearby, rusting away in the trees. Most of the time the road appears as a bear trail in the woods and is very difficult to find beyond Eagle Creek.

The boulders decrease in size as you approach Eagle Creek, a clear-running stream that can be forded on foot during most of the summer. Camping is only fair near the mouth of Eagle Creek; it's well worth trekking another 2 miles to spend the night at Echo Creek, a much more pleasant spot to pitch a tent. The 2 miles to Echo consist of shingle beach the first mile and fine gravel the second. After passing the mouth of Eagle, you cannot spot the stream from the beach even though it parallels the ocean, as a line of trees blocks it from view.

The area near the mouth of Echo Creek is an inviting place to camp and ideal for the first night from Lituya Bay. To the south the shoreline curves and gives way to a view of La Chaussee Spit. To the north is a straight line of beach with one set of waves rolling in after another. The cold, clear creek features excellent fishing for Dolly Varden and cutthroat trout and during the late-summer spawning runs, pink salmon. The place is well supplied with driftwood and rounded rocks for a backcountry sauna, and the nearby grassy benches are covered with strawberry plants. Those parties hiking south find the mouth of Echo Creek a refreshing place to spend any extra days not already claimed by foul weather.

The remaining 3.5 miles to Justice Creek is a hike north along a sandy, sometimes shingle, beach. The beachcombing is excellent in this stretch as there are crab pots half-buried in the sand from one end of the beach to the other, along with mounds of fishing net. Hikers will also see hundreds of dungeness crabs washed up by the powerful waves and quickly devoured by seagulls, eagles, or bears that leave the remains scattered in the sand. Occasionally a lucky hiker will be the first to come upon a large crab struggling back to the ocean, and end up with a memorable dinner. There's no water on this stretch until you reach Justice Creek, spotted by the definite break in the spruce trees on the grassy terrace above the beach.

Fairweather Glacier
Justice Creek to Fairweather Glacier Slough
Distance: 5 miles

At Justice Creek the best camping spots are in the grassy meadow high above the beach, where the stream takes a sharp swing to the east to parallel the ocean for a way. Here is another haven for strawberry lovers. Within the deep pools of the creek lie cutthroat trout waiting to be enticed by a spinner or fly. Near the bend of the creek, hidden in a growth of alder, are the remains of the old miner's cabin.

Justice Creek picked up its name from a trial and hanging that took place here in January, 1899, an event made famous by Jack London in his story "The Unexpected." The summer before, Edith and Hans Nelson and three single men arrived and prospected the area. They found some $8,000 worth of gold dust before extreme temperatures and snow trapped them there for the winter. A lack of food combined with small quarters was apparently an emotional strain for the five miners. One day Michael Dennin shot and killed the two other single men. Hans Nelson overpowered him, however, and the couple was then faced with the problem of justice in a place that was hundreds of miles from the nearest judge and court. In the dead of winter the Nelsons decided to hold their own trial—serving as judge, witness, and jury—and found Dennin guilty. The man himself confessed to wanting everybody's share of gold, and the Nelsons found no alternative but to hang him.

Justice Creek flows north for a spell before emptying into the ocean. The hiking here is along a sandy beach covered with small rounded stones; many of them white quartz that are so smooth they appear as eggs in the sand. There will also be tracks criss-crossing everywhere—the prints of brown bears, eagles, and wolves, but very few, if any, of people. Brown bears are frequent visitors to the shoreline, and often at dusk you can sit on the grassy cliffs overlooking the beach and watch one amble along a half-mile away, as the huge mammal searches for something to eat.

The shoreline remains unbroken for 3 miles north of Justice Creek as you trek toward Cape Fairweather until a clear-water stream is approached. Near its mouth, the stream is jammed by logs and driftwood, which provide a convenient bridge across. Within a mile, rocks begin to appear on the sandy beach; a half-mile farther you are hiking along a shoreline of small rounded stones that make for loose footing, especially when carrying a heavy pack. From here the main river that carries runoff from the Fairweather Glacier is just another half-mile to the north.

The river (marked on topographicals by the benchmark "Arch" and referred to by many as Fairweather Glacier Slough) is actually a wide, fast-flowing waterway that is murky from the large amounts of suspended glacial silt. A raft is needed to ferry people and equipment across. The river changes its character, constantly shifting channels, so backpackers have to study it carefully before selecting where to cross. Usually it is best to avoid the mouth and to hike a quarter-mile or so upstream to wider sections, where quiet eddies might be found along the side. Heavy brush makes it impractical to hike inland in an attempt to ford the river on foot. Such a hike might be possible, however, as a side trip (without your packs) to view the glacier.

Good campsites exist nearby, especially on the north side of the river's mouth, where steep, sandy banks lead up to a flat area that is heavily littered with driftwood and whole trees. Be cautious near the banks, as they are undercut by the glacial water, making them unstable and liable to suddenly collapse into the rushing water.

Traversing Cape Fairweather
From Fairweather Glacier Slough to "Fair" Benchmark
Distance: 10 miles

This is by far the most difficult day of an already challenging trek. For almost 5 miles continuously, you will be trekking across the boulders that make up the rugged shoreline around Cape Fairweather. Hiking across such a coastline is extremely tiring on the ankles and calves and can be endured only by starting early in the day and taking plenty of rests. The terrain is intriguing as the constant pounding of waves has shaped and sculptured the large rocks, some 10 to 15 feet tall, into unusual formations with smooth sides and gentle curves.

Immediately north of the Fairweather Glacier Slough the beach consists of sand and scattered boulders, with a backdrop of spruce-covered cliff. The shoreline seems to be eroding in this stretch; for the next mile the number of spruce trees that have tumbled down from the cliff is impressive. The cliffs themselves add to the unusual atmosphere of this stretch as they echo the pounding waves, making the surf sound even more awesome.

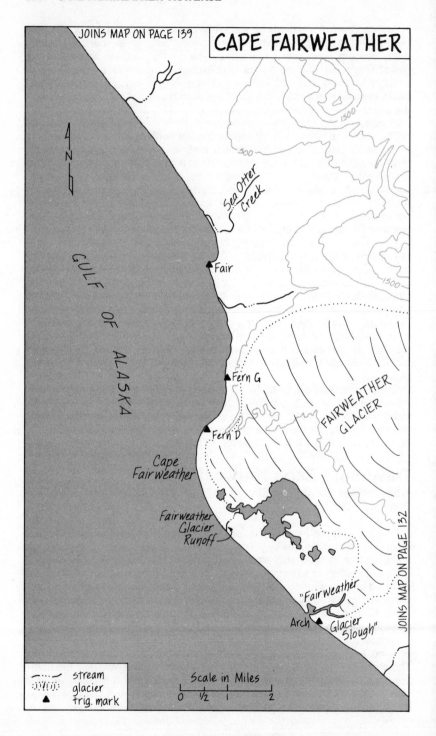

CAPE FAIRWEATHER

JOINS MAP ON PAGE 139

JOINS MAP ON PAGE 132

Sea Otter Creek

GULF OF ALASKA

Fair

Fern G

Fern D

Cape Fairweather

FAIRWEATHER GLACIER

Fairweather Glacier Runoff

"Fairweather Glacier Slough"

Arch

N

1500

500

1500

stream
glacier
trig. mark

Scale in Miles

0   ½   1   2

A mile from the slough the sand ends and you begin bouldering around Cape Fairweather. The hiking is time consuming; backpackers should plan on taking 45 minutes to more than an hour to cover a mile. If possible, attempt it during low tide, but avoid the lower section of the shoreline where rocks combine with seaweed to create slippery footing. Within 1.5 miles, the washed-up trees along the rocky shore increase significantly, and soon you arrive at the second major runoff stream from the Fairweather Glacier.

This is a scenic spot and a good place for an extended break or lunch. The ice-cold water roars out of the opening of spruce trees, but the mouth of the river is jammed by hundreds of trees from the 1958 Lituya Bay landslide and tidal wave. It is possible to cross the stream, skirting from one log to the next, or to sit on the trunk of one and dip your overworked feet into the glacial current to revive them.

From the northern runoff you continue to boulder, and in the next 2 miles you'll round the head of Cape Fairweather. The coastlines north and south of the cape remain hidden, and your view is restricted to a vast stretch of ocean to the west. Several small streams tumble through the rocks and boulders, and the most unusual rock formations, sculptured by the pounding surf, are seen here. Gradually sections of the outer coast appear to the north. After 2 miles you'll be able to see a small strip of the Grand Plateau Glacier to the northeast.

For the first time in over 3 miles you get a short reprieve from bouldering. A shallow cove with distinctive purple sand appears 4.5 miles from the slough (beginning near the trig mark "Fern D" on the topographicals). After you skirt the cove, the boulders return for the next mile as you round a small point (trig mark "Fern G"). An alternative to the rugged coastline at this stretch is to move inland and look for a bear trail 50 to 100 yards within the spruce trees. It is debatable which route is easier or quicker.

Once around the point, a coastline of sand, soft to the feet and easy on the legs, returns. For many this is a glorious moment after a long and tiring day, a time to kick off the boots and let the sand massage and soothe sore foot muscles. In a mile, you'll reach a small, clear stream—a pleasant campsite. Another mile farther north is a distinct point with a rocky coastline, followed by a small bay (trig mark "Fair"). This is a scenic place, with more purple sand and some interesting tidal pools to explore, one of the few you will come across on this route. If planning to camp here, pick up water at the preceding stream as the stream here is usually dry.

Sea Otter Creek
From "Fair" trig mark to first Grand Plateau Glacier stream
Distance: 8 miles

North of the small bay, more sandy beach makes for pleasant hiking, as you reach the shifting mouth of Sea Otter Creek within a mile. The stream originates several miles to the east at Sea Otter Glacier and flows parallel to the shoreline for a half-mile to the north before emptying into the ocean. The current is strong, but backpackers often ford it on foot. The stream that topographicals show to be nearby is often dry through much of the summer.

The trek north continues on sandy beach scattered with big logs. Within 3.5 miles you encounter another large glacial stream that curves out of the trees and parallels the shoreline for a short way before emptying into the Gulf of Alaska. This stream can be considerably wider and deeper than Sea Otter Creek and during times of high water may require rafting people and gear across. Whether rafting or fording on foot, search its banks upstream for the best possible place to cross. The stream is a gathering of several branches flowing from Desolation Valley and Grand Plateau Glacier and almost certainly changes in character from one year to the next.

Once across the fine beach, you walk 3.5 miles or so to the first meltwater stream from the lakes in front of Grand Plateau Glacier. Usually there is little opportunity to gather water along this stretch.

Grand Plateau Glacier
First runoff stream to Clear Creek
Distance: 11 miles

The first runoff stream from the lakes in front of Grand Plateau Glacier swings to the northwest and parallels the shoreline for a half-mile before emptying into the ocean. Good camping exists anywhere along this stretch. The USGS maps show a second stream less than a mile to the northwest, but there is a good possibility this one might be dry. The trek resumes as pleasant beach walking for the next 4 miles, with little or no water encountered. Eventually the forested ridge closes in on the shoreline and the sandy beach turns to a stretch of boulders.

From here it is about a mile to the major runoff of the Grand Plateau Glacier along a shoreline of large boulders that makes for difficult hiking conditions. A better alternative is to search for a bear trail that cuts up the ridge where the sandy shoreline meets the first set of boulders. The trail is hard to spot at first, but once found it appears as a definite path through the lush spruce forest, with an occasional pile of bear scat along the side.

Keep in mind that bear trails can be confusing. Avoid paths that veer inland; instead, continue to skirt the 100-foot ridge that hugs the shoreline. Most likely there will be a trail that stays just above the shoreline and might even afford an occasional glimpse of the water. Somewhere along the line you will drop over what might appear as a dry stream bed or, if it has rained lately, a wet area that appears on USGS maps as a series of small ponds. Most bear trails here curve inland toward the major outlet of the glacial lake.

The main outwash stream of the Grand Plateau Glacier is a deep torrent that splits into two mouths where it empties into the ocean. To attempt to ford the river along the shoreline is extremely dangerous, if not impossible. The safest way is to bushwhack a half-mile east toward the glacier and paddle across the lake near the beginning of the short river. Hacking your way to the lakeshore is not an easy task but will end with a spectacular view of the glacier to the east. Even attempting to raft at the river's wide section in the middle is risky because of the fast currents sweeping down toward the ocean.

Once on the other side of the lake near the east end of the river, it is just under a mile or so northwest to the shoreline, where the boulders give way to sandy beach. Again, bear trails can be sought and followed most of the way to make the hike through the forest considerably easier. The key is locating the right one.

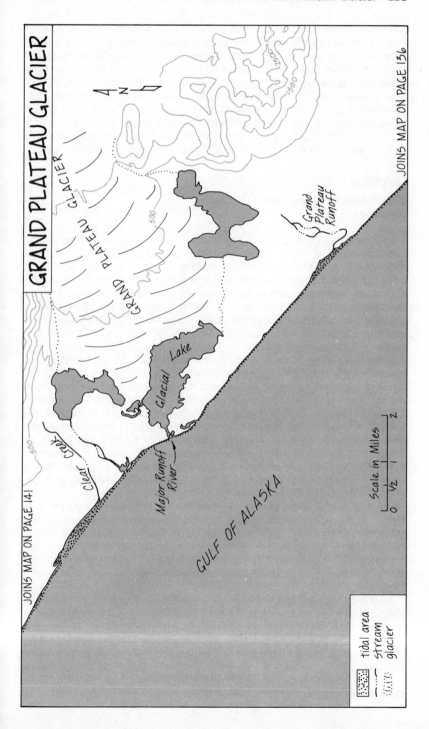

GRAND PLATEAU GLACIER

JOINS MAP ON PAGE 136

JOINS MAP ON PAGE 141

GRAND PLATEAU GLACIER

Grand Plateau Runoff

Clear Creek

Glacial Lake

Major Runoff River

GULF OF ALASKA

Scale in Miles

0   ½   1   2

tidal area
stream
glacier

A return to the sandy shoreline is usually a relief after the last few miles of thick brush, devil's club, and bear trails that end suddenly. After a mile of pleasant beach walking, you come to the last stream issuing from Grand Plateau's glacial lakes. Most likely it no longer empties into the ocean, but rather goes underground 100 yards short of it. If you hike near the surf you'll miss it, as you need to scramble up a small grassy hill to view it. In another 1.5 miles, Clear Creek begins paralleling the shoreline for a spell before gradually cutting toward the sea. The mouth of the creek is not reached until 3 or 4 miles from where the shoreline of boulders from Grand Plateau Glacier returns to sand. There is good camping in areas near the creek.

Dry Bay
Clear Creek to Alsek River
Distance: 13 miles

The hike from Clear Creek to Dry Bay is along a sandy shoreline so soft that it tires the feet and ankles out almost as quickly as a stretch of boulders does. You'll soon discover the best walking is near the water, where the sand is harder, but you need to keep a watchful eye on the surf breaking across the beach. The terrain also changes beyond Clear Creek, with the shoreline gradually becoming a wide, sandy beach and the trees thinning out until they disappear altogether after Doame River. Campsites here are exposed to winds, so tents should be well secured and even reinforced with rocks.

Before departing from the mouth of Clear Creek, fill your water containers, for it is a dry 4.5 miles to the Doame River. The beachcombing is interesting along this stretch, as you encounter everything from crab pots, bottles, and orange floats, to shells, driftwood, and an assortment of crabs. A variety of tracks, including eagle, wolf, coyote, and bear, criss-cross the sand in every direction. A guidebook to wildlife tracks is handy on this trip, although this far north hikers should be ready for some tracks made by three-wheel, all-terrain vehicles. Such machines are popular with fishermen in Dry Bay who occasionally drive them south along the shoreline. This is an illegal activity in the national park, and the park service has begun to enforce the regulation.

You probably won't be able to see the mouth of Doame River, as the river drains underground into the ocean. The river itself can first be seen some 4 miles from Clear Creek, but you have to scramble away from the shoreline through the beach grass to the east to spot it. In another mile, you will spot the island with a cabin on it, which on USGS maps is shown to be near the mouth of the river. Again, hikers who stay near the surf will miss it altogether.

After crossing the Doame River you enter Glacier Bay National Preserve, which offers the same protection for the *land* as in the park, except that in the preserve hunting and commercial fishing are allowed. The cabins viewed on the final stretch are built on land assigned to commercial fishermen. These cabins are private property, so all hikers should respect them as such.

The trek continues for the next 7.5 miles as pleasant beach walking. On a clear day there are magnificent views of the Fairweather Range, and Mount

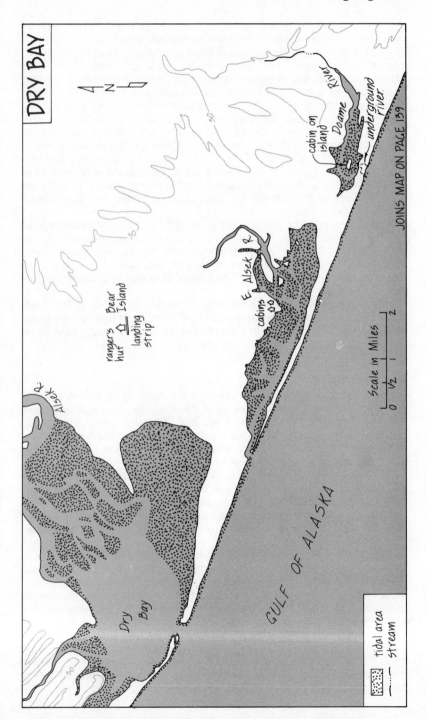

DRY BAY

N

cabin on island

Doame River

underground River

E. Alsek R.

ranger's hut

Bear Island

landing strip

cabins

Alsek R.

Dry Bay

GULF OF ALASKA

Scale in Miles

0   ½   1   2

JOINS MAP ON PAGE 139

tidal area

stream

Fairweather itself, to the southeast. No streams are found (in fact, there is little water of any kind) until the shoreline gradually turns into a sandy spit 4 miles from the Doame River island. A short scramble across brings you into view of the East Alsek River, and the mouth is reached 13 miles from Clear Creek. It is best to ford the river at low tide, even then it could be 3 to 4 feet deep. To cross without a raft would require hiking upstream until the river begins to braid. There are also a few fishermen on the river; occasionally you can arrange for a short lift across.

Along the north side of the East Alsek River you will see a number of small cabins and huts along with various dories and small fishing boats. Narrow jeep tracks wind across the flat, dusty delta that surrounds Dry Bay and lead in a northerly direction toward the Alsek River. Most of them pass the major airstrip at Bear Island, some 4 miles inland, and then run by the green building of the fish-storage company that serves the fishermen of Dry Bay.

Radio contact may be possible at the fish plant, but there are state regulations that prohibit visitors from entering the building. Supplies are generally not available to hikers. In the past, the National Park Service ranger station was located 2 miles up the Alsek River, but plans call for it to be moved closer to the fish plant in 1987. The ranger here can help hikers confirm pick-ups by radio.

The tractor trails throughout this large delta and up the Alsek River are unmarked and confusing as they no longer correspond exactly with those on USGS maps. It is best to ask directions from residents along East Alsek River or people in three-wheelers passing by.

ALSEK BAY

Some 10 miles to the east of Dry Bay, the Alsek River makes a sharp swing to the north in front of the face of Alsek Glacier. This wide section of the river, commonly referred to as Alsek Bay, is another spectacular spot in this land of the glaciers. To the east is the 7-mile face of the Alsek Glacier, to the west the Brabazon Range and, to the south, overshadowing everything the summit of Mount Fairweather.

The glacier is an active one, rumbling throughout the night and turning the bay into a sea of ice. In the middle is Gateway Knob, an island that sits a mile from the immense face. Here you can camp along a shoreline of driftwood and ice, watch huge bergs float by against the backdrop of the Alsek's towering face, and study a skyline of ragged peaks in the Fairweather Range.

This is one of Glacier Bay's more precious gems, a remote gathering of ice, rock, and rugged landscape. Commercial fishermen in Dry Bay occasionally talk of reaching the edge of Alsek Bay by traveling along the south bank of the river, where in the beginning there are tractor trails. But for the most part, the several hundred visitors who reach Alsek Bay every summer are rafters. The 125-mile float along the Tatshenshini and Alsek rivers, beginning in Canada's Yukon Territory and ending at Dry Bay, is one of the most impressive floats in North America. The two rivers provide the only break in the coastal mountains from Cape Spencer to Copper River, and at their confluence, the water flow is three times that of the Colorado River in the Grand Canyon.

12 RAMBLING IN THE FAIRWEATHER RANGE

by Ken Leghorn

Peak 5336 has no name other than its height. It lies somewhere within the Fairweather Range among dozens of other unnamed peaks, accessible after an exhilarating hike from the shore. Reaching its summit in early summer, I drop my backpack and look in amazement at the scene below, forgetting for a moment the arduous ascent. There is something in heights that lends perspective, and looking down I can clearly imagine the ice-bound history of Glacier Bay.

From this spot between sea and sky it is easy to envision the land as it emerged from the last great ice age thousands of years ago. Looking out horizontally, I see black and gray jagged peaks, all equally rugged, rising in every direction. Many glaciers are visible, yet they seem like small shreds of the massive valley glaciers I view at tidewater. It's as if flecks of ice have been thrown randomly among the peaks and left lying at steep angles, filling every pocket and groove. But soon the illusion becomes apparent—each small, hanging glacier is really a solid mass of blue ice 100 feet or more thick, placing incalculable weight on the rock surfaces beneath.

The sharp ridge below me gives way to rounder contours. Below 4,000 feet the country looks worn and polished, and a total picture is formed. My eyes fill the fiords and valleys of Glacier Bay with ice reaching a height of several thousand feet, as it was during the Ice Age. The advance and subsequent retreat of this ice field have ground and smoothed the mountains underneath, carrying millions of tons of sediment into the sea. But above the ice, the high mountains have retained their angular edges.

Now the mighty glaciers lie in only a few major valleys. The rest of the ice has withdrawn into the high peaks, leaving frozen fingerlings protruding from each retreat as if testing the air, waiting for the right temperature and moisture conditions to begin another steady advance down the valley sides.

Just as paddling from the entrance of the park into tidewater glacier country is a reverse lesson in postglacial plant succession, so is climbing from its shores into the alpine zone a trip back through each geologic stage of the Ice Age. Not much matters on peak "5336" except the constant orchestration of rock, snow, rain, and ice. The timelessness and rugged peacefulness I feel here add perspective to many things and coax me to return to explore, climb, and sit awhile on top another unnamed summit.

A ROCK AND ICE WILDERNESS

Indeed, a lifetime of rewarding explorations awaits the adventurous backpacker or mountaineer inside the park. Wherever you go, nature remains the dominant presence, and if some crusty prospector, eager glaciologist, or wilderness-seeking mountaineer has preceded you years ago, their traces probably have long since vanished.

There are no trails in Glacier Bay National Park. The route you pick may have never been traversed; the mountain you climb might have felt the tread of people once before—or never. There are no well-traveled access routes into the mountains, and yet access is virtually unlimited provided you have the necessary skills, equipment, and endurance to venture away—and up—from the protected shores.

The emphasis is definitely on the word *up,* for there are few low-lying valleys or flatlands in Glacier Bay. Although the places where rocky shore and meadow meet ice-studded waters are exquisitely beautiful, it is only when you travel farther back from the steep shoreline of the fiords that you can feel a deep appreciation for the infinite wilderness of the interior peaks.

The mountains themselves do not form a single range, but rise everywhere throughout the park. Lying between the Gulf of Alaska and the inside bay are the better-known peaks of the Fairweather Range. Mostly over 10,000 feet high, these mountains climb out of saltwater on two sides to exhibit some of the greatest vertical relief in the world. The names of these mountains, starting in the south and running west, read like a roll call of past explorers: LaPerouse, Dagelet, Crillon, Bertha, Lituya, Salisbury, and finally, 15,300-foot Mount Fairweather. Ascending any of these glaciated peaks is a major mountaineering effort, especially given the vagaries of weather rolling in off the Gulf of Alaska onto these slopes.

Along the eastern edge of Glacier Bay lies the Chilkat Range, a lower-altitude and gentler set of peaks and high ridges. Having been greatly eroded down to a more human scale during the last glaciation, the Chilkats make excellent alpine hiking and climbing if you can persevere through the bottom 2,000 feet of dense summer brush.

In between the two ranges, dozens of unnamed and largely unexplored mountains rise out of the inside waters of Glacier Bay. Some of these climbs, known only by their height, make excellent one-day scrambles from tidewater if you use sturdy boots and an ice axe. Others require several days of strenuous backpacking up glacier valleys and along scree-covered side hills, bringing you into a mountain world few have known.

From the seat of a kayak, far below, the avid mountain dreamer will spend many moments scouting routes on these peaks, longing for the next opportunity to journey upward.

VENTURING UPWARD

Route is too precise a term. The following are best described as *access possibilities* into the mountains that surround Glacier Bay. This chapter is aimed at self-reliant and creative backcountry travelers who need only to be pointed into the general direction and for whom the freedom of discovery and exploration is their greatest passion. The information is meant to be illustrative, not comprehensive. Even a lifetime exploring the park only proves that the Fairweathers and surrounding peaks can be approached in as many ways as there are inches on a map.

Within the descriptions, *mountaineering* refers to trips that involve glacial travel, snowfields, steep ridges, and elevation gains from sea level to 4,000 feet and higher. Standard equipment for these trips during the summer includes ropes, heavy boots with crampons, ice axes, ice screws and/or snow pickets, and other basic snow equipment and clothing. The term *backpacking* refers to trekking in areas more easily reached in one to several days, areas mostly under 4,000 feet, where a minimum of snow and ice is encountered. All backcountry users should be aware that the snow line throughout much of Glacier Bay lies under 2,000 feet until early July, and that in August many slopes and gullies are filled with snow. You should carry an ice axe on all overnight alpine excursions.

There are formal check-in procedures with the National Park Service; all backpacking and climbing parties should register their trips with the chief ranger at Bartlett Cove. Climbing parties should also keep in mind that they have to be totally self-reliant in the mountains, as outfitting and guiding services in the park are restricted to water-based and shoreline activities.

Outer Coast: from Taylor Bay to Dry Bay

Taylor Bay and Brady Glacier off of Cross Sound provide relatively straightforward access onto Brady Glacier, the perpetually frozen plain that lies at 3,000 feet and extends 40 miles north from its tidewater bay, across the foot of the Fairweather Range, returning to the sea at Reid and Lamplugh glaciers. Ascending Brady Glacier brings fantastic ski touring for the seasoned glacier skier and telemarkskier, as well as access to the eastern bases of Mounts LaPerouse, Crillon, Bertha, and many smaller peaks.

Departing from Cross Sound and continuing north along the outer coast, you encounter a long section of beach from Icy Point to Lituya Bay, broken up by several medium-sized creek outwashes and the face of LaPerouse Glacier. The beach in front of Finger Glacier was the starting point for a successful climb of Mount LaPerouse in 1953, and undoubtedly numerous other inland explorations can be made from points along this coast for those willing to thrash through the dense brush near shore.

Historically, Lituya Bay provided access into the Fairweathers, although the traditional route up Desolation Valley from the head of the bay has been abandoned by most parties who choose instead to travel Fairweather Glacier. But the right history of Lituya Bay and its dramatic head wall make this a magnificent site from which to base future, less summit-oriented climbing ventures.

There is really only one route of choice to reach the Fairweather Glacier from the outer coast, and that follows the north side of a small slough draining the southern lobe of the ice (*not* the creek draining the lake in front of the glacier). Once on the Fairweather Glacier you can ascend for miles along easy moraines and low-angle ice, arriving at stupendous ski, snowshoe, and climbing country past the junction with Desolation Valley.

Dry Bay and Surrounding Area

It takes a lot of imagination and maybe a bit of craziness to plan a climbing trip out of Dry Bay. Yet begin a dozen miles south of it or go up the Alsek River 10 miles to Alsek Bay, and the possibilities are exciting. Both the Grand Plateau and Alsek glaciers offer direct route access to some very broad, smooth, snow-covered glaciers leading up to the boundary peaks of Mounts Lodge, Root, and Fairweather, along with countless lower-altitude nunataks. Routes onto these glaciers should be well scouted, with ample time allowed for negotiating the more heavily-crevassed lower stretches before reaching the land of endless ski mountaineering above.

Rarely does a party of rafters make the beautiful journey down the Tatshenshini and Alsek rivers to Dry Bay without wishing they had packed their skis, crampons, ropes, and backpacks... along with an extra month's worth of food. The peaks above these rivers are enticing beyond description and seldom, if ever, scaled. The Melburn, Grand Pacific, and Alsek glaciers can all be reached from riverside and provide backdoor entry into the heart of the Glacier Bay mountains.

From Haines South along Lynn Canal

Glacier Bay from the east is well protected by the rugged Takhinsha Mountains and the brush-covered Chilkat Range. Several glaciers cut through these ranges, however, allowing the adventurer to make trips from Lynn Canal to Muir Inlet on the east side of the mountains. The most obvious routes go up the Le Blondeau Glacier from the Tsirku River valley due west of Haines, and up the Davidson Glacier from the shores of Lynn Canal south of Haines. Backpacking trips can also be done along the eastern edge of the park in the southern Chilkat Range by hiking from Lynn Canal or Excursion Inlet. Although you may encounter horrendous brush at lower elevations, especially in late summer, you'll find that spending a week or so along the splendid interconnecting alpine ridges makes the struggle well worthwhile.

West Arm

A variety of backpacking and mountaineering trips are feasible from the waters of Johns Hopkins and Tarr inlets. The ridges on both sides of Reid Inlet and along Lamplugh Glacier are wonderful avenues for mountaineers venturing back to the bases of high Fairweather peaks. Several high glacier valleys can be reached on the north side of Johns Hopkins Inlet and there are similar valleys and ridges leading up from Tarr Inlet. Farther south in the West Arm, the shores of other small inlets all offer possibilities for the hearty backpacker or climber.

You will find no route descriptions for any of these trips, unless perhaps in an old prospector's diary, for most modern visitors to the inland waters of the West Arm seldom venture much beyond the shoreline.

EPILOGUE

Our food was almost gone, a week of intense May sun had softened the deep winter snows into waist-high slush, and the gleaming summit dome of Mount Fairweather seemed higher above us now than ever before. I cursed myself for having suggested we try taking a "shortcut" around the second ice fall by ascending the steep slope on its flank. This route had led us onto an endless series of gullies and ridges, ultimately causing a three-day delay. The sun, meanwhile, continued to melt the snow bridges across the crevasses we still had to maneuver through in the days ahead.

Two weeks later, back on the black sand beach of the outer coast, feeling hungry and defeated, my partner and I reflected on the incredible country we just had the privilege of seeing. The disappointment of not reaching the top had already faded, becoming replaced with memories of powdered ski runs, the glint of the full moon on icy slopes, the cries of migrating sandhill cranes as they skirted the glacier and continued north.

Later that night we watched from our campfire as two black wolves danced in the reflections of the moonbeams on the water. As I drifted off to sleep, I could hear the surf giving a thunderous applause and wondered if it was for the wolves or for us having journeyed here to observe them.

It is not a coincidence that the access possibilities described begin and end at tidewater. Like the first mountaineering explorers who arrived over a century ago, climbers for the most part still explore the Fairweathers beginning at sea level. Attempts are being made to preserve the original remoteness of this range—access by ski plane or helicopter is not allowed in some sections of the park, and strongly discouraged in the rest. In an age when success is measured by first-ascent variations, when ski planes leapfrog climbers halfway up a peak so they can begin their dash to the summit, when in the end it is the destination rather than the journey that is noted; the Fairweathers offer a different set of values.

Mountaineers can go elsewhere to climb higher and can undoubtedly stay at home to find better rock. But the few that venture into the Fairweathers and surrounding peaks experience the timelessness of high glaciers and mountains scarcely altered.

There is a purity in mountaineering, one that reflects the rugged, reborn land of Glacier Bay.

AFTERWORD:
BREAKFAST NO. 10

Breakfast on the final day of my 1986 expedition to Glacier Bay: Bill suddenly appears at the door of the tent and says, "Surprise, coffee in bed."

That's it; a cup of instant coffee. No oatmeal, no granola, nothing else; just coffee that is lukewarm at best. But it is a pleasant surprise.

Last night we were sure we had used up the fuel cooking our final dinner in Glacier Bay. We were heating some water for two pouches of tuna neptune when the stove sputtered and died. The water wasn't even close to boiling. We poured the warm water on the freeze-dried dinners anyway and ate noodles and chunks of tuna that crunched like trail mix.

But hunger spurs ingenuity, and on the morning of our pick-up Bill somehow coaxed a little more flame from the remaining tablespoon of gas in the bottle. What a pleasant ending. On a cloudless morning in the West Arm, I now find myself lying in a warm sleeping bag, enjoying an unexpected cup of coffee while staring out at Reid Glacier.

All of this releases a flood of emotions as I contemplate the inevitable; that soon a floatplane will thunder in and whisk us away. I first arrived in 1978 but in the last two years I've given this place my full attention; reading about it in the winter, studying topographicals in the spring, exploring it in the summer. And now it's over. This book will be written and I will look elsewhere for adventures.

But on this last morning, Glacier Bay again unveils its beauty to me like it has so often before. The sun has just appeared over the mountains, turning Reid Glacier and its inlet of ice into a sparkling, almost blinding, panorama. Through the door of a dome tent, I look over this wondrous scene and hope for two things. That someday I can return with my children. And that when I do, the cruise ships, sightseeing planes, sailboats, and, yes, campers, hikers, and kayakers, have not altered this priceless wilderness.

Glacier Bay will always be changing; I just hope man serves only as a witness and never as the cause.

Time to get out of my bag and move on. The hum of the floatplane is off in the distance.

APPENDIX

Air taxi and charter services that provide pick-up and drop-off transportation into the backcountry of Glacier Bay National Park are listed below:

JUNEAU

Channel Flying, Inc.
2601 Channel Dr.
Juneau, AK 99801
Tel. (907) 586-3331

Wings of Alaska
1873 Shell Simmons Dr.
Suite 119
Juneau, AK 99801
Tel. (907) 789-0790

GUSTAVUS

Glacier Bay Airways
P.O. Box 1
Gustavus, AK 99826
Tel. (907) 697-2249

SITKA

Bellair
P.O. Box 371
Sitka, AK 99835
Tel. (907) 747-3220

Mountain Aviation
P.O. Box 875
Sitka, AK 99835
Tel. (907) 966-2288

SKAGWAY

Skagway Air Service
P.O. Box 357
Skagway, AK 99840
Tel. (907) 983-2218

HAINES

L.A.B. Flying Service
P.O. Box 272
Haines, AK 99827
Tel. (907) 766-2222

YAKUTAT

Gulf Air Taxi
P.O. Box 367
Yakutat, AK 99689
Tel. (907) 784-3240

INDEX